£23-50

International Studies in Educational Achievement
VOLUME 1

The IEA Study of Mathematics I: Analysis of Mathematics Curricula

The IEA Study of Mathematics I: Analysis of Mathematics Curricula

KENNETH J. TRAVERS

and

IAN WESTBURY

University of Illinois at Urbana-Champaign, USA

with assistance from

Elizabeth E. Oldham
H. Howard Russell
Robert A. Garden
James J. Hirstein
A. I. Weinzweig
Richard G. Wolfe
Ian D. Livingstone

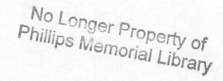
Published for the International Association
for Educational Achievement by

PERGAMON PRESS

OXFORD · NEW YORK · BEIJING · FRANKFURT
SÃO PAULO · SYDNEY · TOKYO · TORONTO

QA
11
I32
1989

U.K.	Pergamon Press plc, Headington Hill Hall, Oxford OX3 0BW, England
U.S.A.	Pergamon Press, Inc., Maxwell House, Fairview Park, Elmsford, New York 10523, U.S.A.
PEOPLE'S REPUBLIC OF CHINA	Pergamon Press, Room 4037, Qianmen Hotel, Beijing, People's Republic of China
FEDERAL REPUBLIC OF GERMANY	Pergamon Press GmbH, Hammerweg 6, D-6242 Kronberg, Federal Republic of Germany
BRAZIL	Pergamon Editora Ltda, Rua Eça de Queiros, 346, CEP 04011, Paraiso, São Paulo, Brazil
AUSTRALIA	Pergamon Press Australia Pty Ltd., P.O. Box 544, Potts Point, N.S.W. 2011, Australia
JAPAN	Pergamon Press, 5th Floor, Matsuoka Central Building, 1-7-1 Nishishinjuku, Shinjuku-ku, Tokyo 160, Japan
CANADA	Pergamon Press Canada Ltd, Suite No. 271, 253 College Street, Toronto, Ontario, Canada M5T 1R5

Copyright © 1989 I.E.A.

First edition 1989

Library of Congress Cataloging in Publication Data

The IEA study of mathematics.
(International studies in educational achievement ; v. 1)
Contents: 1. Analysis of mathematics curricula.
1. Mathematics—Study and teaching. I. Travers, Kenneth J. II. Westbury, Ian. III. International Association for the Evaluation of Educational Achievement. IV. Series.
QA11.I32 1988   510'.7'1   88-25384

British Library Cataloguing in Publication Data

The IEA study of mathematics. 1.
1. Schools. Curriculum subjects: Mathematics. Academic achievement of students. Assessment
I. Travers, Kenneth, J. II. Westbury, Ian
III. Oldham, Elizabeth E. IV. Series
510'.76

ISBN 0-08-036468-3

Printed in Great Britain by BPCC Wheatons Ltd, Exeter

Foreword

The International Association for Educational Achievement (IEA) was founded in 1959 for the purpose of comparing the educational performance of school students in various countries and systems of education around the world. Its aim was to look at achievement against a wide background of school, home, student and societal factors in order to use the world as an educational laboratory so as to instruct policy makers at all levels about alternatives in educational organization and practice. IEA has grown over the years from a small number of nations and systems of education to a group of over forty. Its studies have covered a wide range of topics, and have contributed to a deeper understanding of education and of the nature of teaching and learning in a variety of school subjects.

This particular study of achievement in mathematics was begun in 1976 and is now complete. This volume is the first in a projected set of three dealing with the subject and is intended to set forth curricular forces behind national and system differences in student performance. It represents a highly arduous enterprise in curriculum analysis and analysis of test results.

The international costs of the study were supported by grants from a number of agencies: The Spencer Foundation, The Ford Foundation, the Deutsche Forschungsgemeinschaft, the National Science Foundation (US), The National Institute of Education (US) and The Center for Statistics of the Office of Educational Research and Improvement of the United States Department of Education. The international coordinating center in New Zealand was also supported extensively by the New Zealand Department of Education, and we are grateful to its Director-General W. L. Renwick. The center at Champaign-Urbana was supported by the College of Education and the Research Board of the University of Illinois at Urbana-Champaign.

This volume represents the efforts of the many people involved in the study to clarify the domain and its assessment. As the editors note, many people were involved in the preparation of the volume. I should like to express the gratitude of IEA and its editorial committee, chaired by Professor Richard Wolf of Teachers College, Columbia University, for the

v

kind offices of the external reviewers: Alan J. Bishop, Cambridge University; M.A. Brimer, then of The University of Hong Kong; and A. Harry Passow of Teachers College, Columbia University. We are also grateful to Pergamon Press, and its publisher, Mr Robert Maxwell, for his kindness in supporting the IEA publications program.

ALAN C. PURVES
Chairman, IEA

Acknowledgments

A project of the scale of the Second International Mathematics Study could not have taken place without the dedication and talent of hundreds of individuals throughout the world. It is inevitable that, in seeking to recognize these efforts, we take a great risk – that we will, by omission or commission, fail to acknowledge every person who helped to bring this volume into being.

First and foremost, we are grateful to the National Research Coordinators, one in each of the participating systems, who directed the data-gathering and data-clarifying task. This was a long and onerous responsibility, and in many cases involved laborious translation. This work was done with the help of National Mathematics Committees, and to those who were involved in these committees we also wish to express our appreciation and thanks for their efforts. The National Research Coordinators were: Malcolm Rosier, Australia; Christiana Brusselmans-Dehairs, Belgium (Flemish); Georges Henry, Belgium (French); David Robitaille, Canada (British Columbia); Leslie McLean, Canada (Ontario); Maritza Jury Sellan, Chile; Michael Cresswell, England and Wales; Erkki Kangasniemi, Finland; Daniel Robin, France; Patrick Griffin, Hong Kong; Julia Szendrei, Hungary; Elizabeth Oldham, Ireland; Arieh Lewy, Israel; Sango Djibril, Ivory Coast; Toshio Sawada, Japan; Robert Dieschbourg, Luxembourg; Hans Pelgrum, The Netherlands; Athol Binns, New Zealand; Wole Falayjo, Nigeria; Gerard Pollock, Scotland; Mats Eklund and P. Simelane, Swaziland; Robert Liljefors, Sweden; Samrerng Boonruangrutana, Thailand; Curtis McKnight, United States.

At the international level, Roy W. Phillipps and, subsequently, Robert A. Garden were the principal coordinators of the study from their offices in the New Zealand Department of Education in Wellington. The members of the International Mathematics Committee provided substantive guidance throughout the project. Members of the International Mathematics Committee were: Kenneth J. Travers (Chairman), United States; Sven Hilding, Sweden; Edward Kifer, United States; Gerard Pollock, Scotland; Hans-Georg Steiner (1976–1979), Federal Republic of Germany; Frederick van der Blij (1980–1983), The Netherlands; Tamás Varga,

Hungary; and James Wilson, United States; A. I. Weinzweig (Consulting Mathematician), United States; Richard Wolfe (Consulting Psychometrician), Canada. During the initial phases of the study, the members of this committee did much of the developmental work, particularly the production of the various working papers, the building of the international grid, and the construction of the international item pools.

Regrettably, one of the members of the committee, Tamás Varga of the Országos Pedagógiai Intézet, Budapest, did not live to see these volumes through to publication. We hope that he would have been pleased with the outcomes of his contributions to SIMS. These volumes will stand as one symbol of Tamás Varga's contributions to mathematics education both in Hungary and on the world stage. Betz Frederick and Vijaya Thalathotti prepared several of the figures included in this volume.

In terms of the present volume, the lion's share of the credit is due to the Curriculum Analysis Group, a task force that was established here at the University of Illinois at Urbana-Champaign. This group undertook the tasks associated with the curriculum analysis, prepared papers for various international conferences, most notably the Osnabrück Symposium (Steiner, 1980) and, of course, saw the present volume through to completion. Elizabeth Oldham of Trinity College, Dublin devoted an enormous amount of continuous, sustained effort to this work. On two separate occasions, Ms Oldham was able to obtain leave from Trinity College to devote several months to the project. Here at the University of Illinois at Urbana-Champaign James J. Hirstein also contributed much to this publication, while A.I. Weinzweig at the University of Illinois at Chicago made numerous trips to Urbana to assist at critical points. H. Howard Russell and Richard G. Wolfe at the Ontario Institute for Studies in Education, Toronto, Canada, were also important team members as the curriculum analysis proceeded. We were extremely fortunate in that at one critical point Ian Livingstone, now Director of the New Zealand Council for Education Research, was able to spend a sabbatical leave in Urbana. His assistance, too, assisted greatly in moving the volume ahead.

In our office, we had support from Judith Ruzicka for computing and data documentation. Several doctoral students also had parts to play in this evolving odyssey, including Gullayah Dhompongsa, Nongnuch Wattanawaha, John Williams, Kazem Salimizadeh, Lynn Juhl and Chantanee Indrasuta. Word processing was done by Del Jervis, Leigh Little, and Denny Arvola. Sally Spaulding ably managed the office.

The manuscript underwent extensive review and revision at several stages. The reviewers include: Alan J. Bishop, Cambridge University, England; Alan Brimer, University of Hong Kong; Leigh Burstein, University of California at Los Angeles, United States; A. Geoffrey Howson, Centre for Mathematical Education, Southampton University, England; Edward Kifer, University of Kentucky, United States; A. Harry

Passow, Teachers College, Columbia University, United States; T. Neville Postlethwaite, Hamburg University, Federal Republic of Germany and Richard Wolf, Teachers College, Columbia University, United States. We thank these individuals for their many helpful comments and suggestions.

A project of the scale of the Second International Mathematics Study could not have taken place without the generous support of the agencies that supported the research. These agencies are listed in the Foreword and we join with Alan Purves, the chair of IEA, in extending our thanks to them. Finally, we must acknowledge the generous assistance offered SIMS by many host institutions – be they Ministries of Education, Research Institutes or Universities – which supported both national and international components of the Study. To all of these agencies, and they are too numerous to name individually, we express our appreciation and gratitude for providing environments within which this endeavor could take place.

This volume has evolved over a long period of time, and in doing so has taken on a somewhat different character at each cycle of the evolutionary process. Accurate attribution of authorship for the chapters is difficult. However, the following played major roles in research and writing throughout the development of this publication:

Chapter 1: Kenneth J. Travers, Elizabeth E. Oldham, Ian D. Livingstone
Chapter 2: Elizabeth E. Oldham, H. Howard Russell, A. I. Weinzweig, Robert A. Garden
Chapter 3: Ian Westbury, Elizabeth E. Oldham, James J. Hirstein
Chapter 4: Kenneth J. Travers, Elizabeth E. Oldham, A. I. Weinzweig, James J. Hirstein
Chapter 5: Ian Westbury, Richard G. Wolfe
Chapter 6: Ian Westbury, H. Howard Russell, Kenneth J. Travers
Chapter 7: Kenneth J. Travers, Ian Westbury

Note to Reader

A supplement to this volume containing the following is available through the ERIC Document Reproduction Center. Descriptions of the school systems participating in both the Population A and Population B components of SIMS; indices of intended coverage for all SIMS items; appropriateness ratings for all items; teacher opportunity-to-learn ratings for all items; and a version of the cognitive item table showing the grid (content and behavioral level) classification of all items, anchor items, i.e., items used in the IEA First International Mathematics Study, and the distribution of items across the cross-sectional and longitudinal studies.

Contents

6. Outputs and Outcomes of Mathematics Education

7. Summary and Implications

List of Tables

List of Figures

Chapter 6

Chapter 7

Abbreviations for Education Systems

Australia	AUS
Belgium (Flemish)	BFL
Belgium (French)	BFR
Canada (British Columbia)	CBC
Canada (Ontario)	CON
England and Wales	ENW
Finland	FIN
France	FRA
Hong Kong	HKO
Hungary	HUN
Ireland	IRE
Israel	ISR
Japan	JPN
Luxembourg	LUX
Netherlands	NTH
New Zealand	NZE
Nigeria	NGE
Scotland	SCO
Swaziland	SWA
Sweden	SWE
Thailand	THA
United States	USA

1
Origins of the Second International Mathematics Study

1.1 Introduction

Throughout the world, mathematics occupies a central place in the school curriculum. In most school systems between 12 and 15 percent of student time is devoted to mathematics. The only other subjects allocated as much time are those associated with language, particularly the mother tongue and reading.

The importance of mathematics in the school curriculum reflects the vital role it plays in contemporary society. At the most basic level, a knowledge of mathematical concepts and techniques is indispensable in commerce, engineering and the sciences. From the individual pupil's point of view, the mastery of school mathematics provides both a basic preparation for adult life and a broad entrée into a vast array of career choices. From a societal perspective, mathematical competence is an essential component in the preparation of a numerate citizenry and it is needed to ensure the continued production of the highly-skilled personnel required by industry, technology and science.

Beyond these practical considerations, it is generally believed that mathematics is fundamental to an effective education. It provides an exemplar of precise, abstract and elegant thought. And while the generalized effects of mathematical studies on overall intellectual development are difficult to analyze, let alone measure, there does appear to be a universal consensus that the study of mathematics helps to broaden and hone intellectual capabilities.

In view of the importance of mathematics in society and in the schools, the efficacy of mathematics teaching and learning demands continued and sustained scrutiny. It is to this end that the study is directed. Its purpose is to compare and contrast, in an international context, the varieties of curricula, instructional practices, and student outcomes (both affective and cognitive) across the schools of twenty countries and educational systems.* By portraying each system's school mathematics programs and attainments against a

---

* "System" is used in this volume to mean educational jurisdiction. For example, since Belgium has, in the context of SIMS, two distinct educational jurisdictions, Flemish and French Belgium are treated separately. (For a list of these systems see Appendix I.)

1

cross-cultural backdrop, each individual system is afforded the opportunity to understand better its own endeavors in mathematics education. Our concern is with what mathematics is intended to be taught, what mathematics is actually taught, how that mathematics is taught, and what mathematics is learned. The results of the adventure in data gathering and analysis is an international portrait of mathematics education—a portrait that we believe adds to our knowledge of the state of mathematics education around the world.

This study was sponsored by the International Association for the Evaluation of Educational Achievement (IEA), a cross-national enterprise devoted to comparative studies of school practices and achievements. In the early 1960s IEA sponsored the First International Mathematics Study (FIMS) (Husén 1967); this study included twelve systems, and of these systems eleven took part in this Study. Because SIMS has drawn on items and questions from FIMS, it is able to present a picture not only of mathematics teaching and learning in the 1980s but also to compare and contrast parts of the picture which has emerged with the findings from FIMS. Thus for the first time, it is possible to draw a sketch of change in mathematics teaching and learning over time in several educational systems.

Following FIMS, IEA sponsored the Six-Subject Survey involving some twenty-one systems, 250,000 students, 50,000 teachers and nearly 10,000 schools. These IEA projects are the most extensive empirical investigations of educational systems ever carried out at the cross-national level and SIMS builds on this experience and expertise. The methods of investigation undertaken within SIMS draw heavily on these earlier studies, but also, we hope, reflect the thinking within IEA following these earlier studies.

This is the first volume of the final report of SIMS. Here we will describe the mathematics curriculum of the participating systems as well as the overall rationale and structure of the study as a whole. Volume 2 will describe the outcomes of instruction in the participating systems. Volume 3 will describe the results of a "longitudinal" study undertaken within a subset of the SIMS systems. The goal in that substudy was to explore the patterns of growth in student learning that take place in the course of a school year and embed those patterns in the context of classroom practices.

1.2   Why a Second International Mathematics Study?

Let us consider each of the aspects of SIMS in turn. It is a *second* study. It is important because of its *international* dimensions. It is an enterprise of major interest to those concerned with and responsible for the teaching and learning of *mathematics* in schools. Finally, it is a *study* (as distinct, say, from a survey).

A SECOND International Mathematics Study

Much has happened since 1964 when FIMS was carried out. In the early 1960s the curriculum reform movement was just getting under way in many countries around the world. Now, curriculum reform, as it was known then, has subsided. What is the current status of mathematics education? To what extent are students receiving adequate mathematical preparation in order to function effectively in today's society? What effect have changes in the patterns and philosophies, such as the move in some systems from selective to comprehensive schooling, had on the teaching and learning of mathematics? What shifts have occurred in the proportions of students enrolling in pre-university mathematics courses? What modifications in curricula to accommodate these changes have occurred? And in all of these areas of investigation, what relationships appear with respect to student attitude and achievement today as compared with those a generation ago?

Another important aspect of a *second* study is that IEA has learned much in the intervening years about the art and science of international survey research. The areas of sampling (both what to do and how to do it), data collection and processing, analysis, reporting, and dissemination have all advanced since the first round of IEA studies, and there has been, fortunately, careful documentation for the benefit of subsequent investigators. Furthermore, considerable training accompanies all IEA investigations and the expertise which has resulted from this is available in the various systems to promote the work of IEA and assist new personnel as new studies are undertaken.

A Second INTERNATIONAL Mathematics Study

In view of all the difficulties introduced by dealing with many different languages, cultures, agencies and administrative authorities, what are the benefits of an international study? Again, this can be answered on several different levels. On the grand scale, it can be argued that whatever differences occur within a system, such as in curricular emphases, these differences will be even greater between systems. Hence, it is of interest to identify patterns of achievement across cultures. We will see, for example, that school geometry varies greatly in content from system to system. One would expect variations in patterns of student achievement to accompany such variations in the content of the curriculum.

A second benefit accrues from the knowledge to be gained from such a project. What are the various patterns of course organization? What positions have been taken with respect, say, to uses of hand-calculators, and for what reasons? How do classroom practices differ across systems and are those differences related to say, the numbers of students who pursue advanced studies of mathematics in school, or achievement?

At the system level, an international project can add important dimensions to each system's assessment activities. In the United States, for example, the National Assessment of Educational Progress (NAEP) conducts periodic mathematics surveys. NAEP, however, focuses on student achievement only, and provides data at only one time in the school year. In the IEA Studies, data on student achievement are presented in the context of information on the content of the curriculum and whether that content was actually taught to those tested. As will be demonstrated in this and successive volumes (Robitaille and Garden 1988; Burstein et al. 1989), the Second International Mathematics Study focuses intensively upon the *context* in which achievement takes place. For example, opportunity-to-learn data were obtained from both teachers and students. For a subset of eight systems (the "longitudinal study") a pretest measure of achievement was obtained and, as a result, careful descriptions were possible of what mathematics was learned during the school year. For these same systems, detailed information was also gathered on how the mathematics was handled by the classroom teacher. The results of this study will be presented in the context of what classroom practices were associated with "high growth" in some systems and not in others, and what dimensions of growth were related to these practices.

On a local basis, an international study can provide a valuable data bank for exploring issues of interest to individual mathematics educators. Examples of such use from FIMS include Postlethwaite's report (1971) on how item information was used by curriculum planners in Sweden and Schildkamp-Kündiger's work (1974) on sex differences in mathematics achievement in West Germany.

A Second International MATHEMATICS Study

In the Second International Mathematics Study, the subject matter of mathematics is fundamental. The International Project Council and the International Mathematics Committee made every attempt to reflect the concerns and priorities of the National Mathematics Committee in each participating system. Systems were in turn urged to appoint national mathematics committees that were aware of national issues in mathematics education and reflected the concerns of the classroom teacher, the mathematics educator at the university level (teacher educator, curriculum developer and researcher), and the professional mathematician.

A Second International Mathematics STUDY

The first important component of this study is the curriculum analysis. Through this study an attempt has been made to provide a context in which to interpret the data obtained from the questionnaires and examinations.

The study also determined to focus upon the *classroom*, what it is that teachers do there, and what are the associated student outputs. The "classroom process" instruments which were developed for the longitudinal study represent a new approach to the study of teaching and learning. Systems and individual scholars were encouraged to pursue lines of inquiry of particular interest to them as the Study developed. The Dutch National Center, for example, has carried out a detailed analysis of the opportunity-to-learn data as it relates to student outcome measures. In the Belgium (Flemish) Center, extensive use was made of the case studies reported in Steiner (1980) as important content issues in the mathematics curriculum were identified and analyzed in the light of international comparative data. Wattanawaha (1986) has undertaken an extensive analysis of the Thai data sets to explore the issue of within-system equality of mathematics teaching and learning.

1.3 Conceptualization of the Study: Three Aspects of the Curriculum

A meeting of mathematicians, mathematics educators and educational researchers was held in 1976, in order to make recommendations as to what SIMS should be like, and what goals such a study should address. This group of some 35 individuals spent five days debating issues, formulating recommendations and preparing reports. As the conference concluded, the concerns of the group emerged in three clusters. One cluster of concerns dealt with the curriculum. As Wilson (1976) put it, in a paper given at that meeting:

. . . the second study should be approached as a curriculum survey. If the study focuses on the content of what is being taught, the relative importance given to various aspects of mathematics, and the student achievement relative to these priorities and content, then the international and national results can help in our understanding of comparable curriculum issues.

A second cluster of concerns had to do with the classroom. There was a determination to obtain information on what teachers do as they teach mathematics. A considerable knowledge base has developed in the study of teaching in a generic sense; there is, however, a need for extensive and detailed information on the teaching of *mathematics*.

The third cluster of these concerns dealt with the end-products of instruction—how much mathematics students learn and what attitudes they have about mathematics. What are the profiles of achievement in plane geometry, or in problem solving? What differences in achievement and attitudes are there between boys and girls? And for those systems taking part in FIMS, what changes in achievement and attitudes have taken place since 1964?

Underpinning the conceptualization of the study is the simple model presented in Figure 1.3.1 and taken up later in an expanded and elaborated form in Figures 1.3.2 and 1.3.3 below. Essentially, it views the mathematics curriculum in a school system as taking on different embodiments at each of three separate levels. At the level of the educational system (the system, the educational region, the school district) there is a set of *intentions* for the curriculum. There are goals and traditions. There are impulses from the community of mathematicians or educators that help shape the character of the curriculum. This collection of intended outcomes, together with course outlines, official syllabi, and textbooks forms an *intended curriculum*. It is what one is told, for example, if the question is asked of an official of the ministry of education. "What do you do in your system about statistics, or the calculus?"

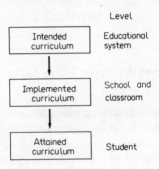

FIG 1.3.1   Framework for the Study: The Three Curriculums

   The second level deals with the classroom, the setting in which the content becomes *implemented* or translated into reality by the teacher. The classroom is central to the educational process, for it is in the classroom, by and large, that children are introduced to the study of mathematics and it is where their concepts and attitudes are formed, and it is the teacher who has the responsibility for transmitting this knowledge to students. As Bloom (1974) has observed: "beautiful curriculum plans have little relevance for education unless they are translated into what happens in the classroom of the nation or the community." No statement of curricular intentions can by itself provide an adequate description of mathematics education in schools. There are important questions about the coverage of subject matter content—the extent to which the prescribed syllabus is actually taught by the classroom teacher. For example, what topics are not taught because they are perceived by teachers to be of little importance? Which topics are seldom reached because of pressures of time?

   Examining patterns of teacher coverage of subject matter—both within and between systems—is an important aspect of curriculum analysis activity and receives a great deal of attention in this volume.

   The third aspect of the curriculum has to do with student *attainment*. After a given period of time at school, the student has acquired a body of mathematical knowledge, and acquired certain attitudes toward the subject and its use in the world. This is the *attained* curriculum. In some ways this aspect of the curriculum is the principal focus of SIMS but in addition to measuring the achievement and the attitudes of the students of the participating systems, SIMS also seeks to address the issue as to what extent these student outcomes can be accounted for by curricular intentions and classroom instruction. Put another way, what aspects of the curriculum as intended, say, by a ministry of education and taught by the teacher, are actually learned by the student?

It should be noted that the model allows for the learning of mathematics which is neither in the intended or the implemented curriculum. For example, topics in vectors might be learned through a computer programming activity studied at home. Or, statistical concepts may be learned through their use in other subject matter areas in school. Indeed the extent to which there is strong correspondence between the three curricula and the degree to which this correspondence differs between systems is an important concern of this Study.

Figure 1.3.1 provides a useful way of viewing the Study, but this figure deals only with matters of mathematical content of the intended curriculum. One must take into account *contextual* factors as one considers, for example, what mathematics is intended for a certain population of students. One such factor is the selectivity of schooling; another is that of the presence or

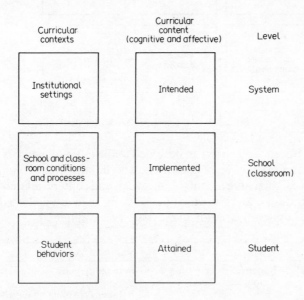

Fig 1.3.2   Contextual Dimensions of the Curriculum

absence of external examinations in a school system. We therefore add to the content dimension of the model a contextual component as seen below.

Finally, there are background variables, or antecedents of the curriculum, that impinge upon both the contextual and content-related aspects of the curriculum (see Figure 1.3.3). For example, at the system level, the wealth of a society may affect the retentivity of the system. Community and home press for school achievement, and for the study of mathematics would be expected to relate to the amount of mathematics taught and learned. The characteristics which the students bring to the class, including prior knowledge of mathematics and their attitudes toward the subject, relate to student classroom behavior and the amount of mathematics known (attained curriculum) at the end of the school year.

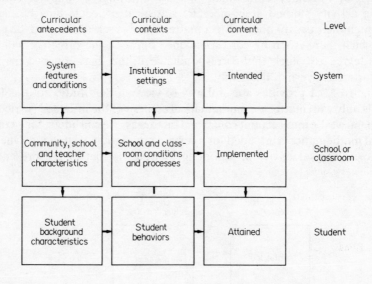

FIG 1.3.3   An Expanded Model for the Study

Such antecedent factors can be added to the left of the matrix to provide a three-by-three depiction of the structure of the study. The columns deal with curricular factors: antecedents, contextual, and content-related factors. The rows indicate the level at which the relevant data are collected and, hence, imply the nature of variation in the data. For example:

	Level	Sources of Variation
1.	System	Between systems
2.	School (classroom)	Between schools
		Between classrooms in these schools
3.	Student	Within classrooms

The model in Figure 1.3.3 also provides a convenient starting point to define the scope of the three volumes in the series of reports outlining the results of SIMS. The present volume, the first in the series, is concerned with variables in the first and second row in the model, the intended and implemented curricula. After a discussion of the rationale for the Second International Mathematics Study, it moves on to examine the characteristics of various systems, the institutional settings in which mathematics learning takes place in each system, and describes some of the differences in the content of the curriculum which have resulted from the various influences that have shaped it in the various systems taking part in the Study.

This volume also presents a detailed analysis of the patterns of "delivered instruction" of mathematics in the various systems, based

Table 1.3.1   *Participation by Educational Systems in Various Aspects of the Second International Mathematics Study*

System	Curriculum analysis	Cross-sectional data		Longitudinal data		Classroom processes data	
		Pop. A	Pop. B	Pop. A	Pop. B	Pop. A	Pop. B
Australia*							
Belgium (Flemish)	x		x	x		x	
Belgium (French)	x	x	x				
Canada (British Columbia)	x		x	x		x	x
Canada (Ontario)	x		x	x		x	x
Dominican Republic†							
England and Wales	x	x	x				
Finland	x	x	x				
France	x			x		x	
Hong Kong	x	x	x				
Hungary	x	x	x				
Ireland†	x						
Israel	x	x	x				
Japan	x		x	x		x	
Luxembourg	x	x					
Netherlands	x	x					
New Zealand	x		x	x		x	
Nigeria	x	x					
Scotland	x	x	x				
Swaziland			x				
Sweden	x	x	x				
Thailand	x		x	x		x	
United States	x			x	x	x	x

Notes: * Australia carried out a Second Study that closely resembled the First International Mathematics Study. The findings are reported in Rosier (1980).
† The Dominican Republic participated in the longitudinal study for Population A. Ireland collected cross-sectional data for Population A in 1986. These countries will publish their own reports on the Study.

upon opportunity-to-learn data from teachers. It concludes with an analysis of the concept of the *yield* of mathematics instruction, the outcome in some senses of the planning and the patterns of delivery of content that are associated with the intended and implemented curricula.

Volume II (Robitaille and Garden 1988) is concerned with the variables in the third row of the model (the attained curriculum) but makes substantial reference to variables in the first column, dealing with characteristics of the national systems, the communities, schools and teachers which have a major bearing on the mathematics learning and the attitudes of students. It describes the results of a cross-sectional study in 20 of the participating systems.

The third and final volume in the series (Burstein et al. 1989) describes the longitudinal study undertaken in eight of the participating systems, and deals again with the implemented curriculum, and, particularly, the classroom processes that form the context for learning mathematics. Table 1.3.1 describes which systems were involved in the varying components of the study.

1.4   Target Populations

The choice of target populations for the Second Study was determined by decisions made at the beginning of the study to parallel the design of FIMS. The FIMS target populations (Husén 1967, Vol. I, p. 46) were:

Population 1a—All pupils who are 13.0–13.11 years old at the date of testing.
Population 1b—All pupils at the grade level where the majority of pupils of age 13.0–13.11 are found.
Population 3—All pupils who are in the grades (forms) of fulltime study in schools from which universities or equivalent institutions of higher learning normally recruit their students. This latter population was divided into two parts:
Population 3a—Those studying mathematics as an integral part of their course for their future training or as part of the pre-university studies, for example, mathematicians, physicists, engineers, biologists, etc., or all those being examined at that level.
Population 3b—Those studying mathematics as a complementary part of their studies, and the remainder.

The basis for the choice of these populations is described in Husén (1967, Vol. I, pp. 46–47). In planning FIMS, the focus was on two "major terminal points" in the school system. The first, the 13-year-old population, was chosen since for some systems taking part in FIMS, this was the last year of universal compulsory fullterm schooling. For the older population, the focus was on securing information on the mathematics achievement of students who had completed pre-university schooling.

In the SIMS, target populations were selected to parallel as closely as possible those in the FIMS. The definitions used for SIMS were as follows:

Population A: All students in the grade in which the modal number of students has attained the age of 13.0–13.11 years by the middle of the school year. (The comparable population for FIMS is Population 1b.)

Note that "majority" was replaced by "modal number," since in some systems, due to wide spreads in ages across grade levels, no age group held a strict majority. It should also be noted that by the time of the Second Study, the criteria of "last year for compulsory schooling" no longer applied. In most countries in the Study, the termination of compulsory school was well beyond 13 years of age.

Population B: All students who are in the normally accepted terminal grade of the secondary education system, and who are studying mathematics as a substantial part (approximately 5 hours per week) of their academic program.

Although the SIMS Population B is targetted at the same level as the FIMS Population 3, it only embraces *part* of the FIMS population, FIMS Population 3a. In SIMS, no study was made of the "nonspecialist" mathematics students (Population 3b in FIMS). The major reasons for not including a target population in SIMS analogous to Population 3b were those of time and resources. Primary evidence from participating countries indicated great variation in curricular offerings for such "nonspecialists". It was decided that it was impractical to undertake the curriculum study and test development activity that would be required to involve this group of students in the project, even though it was recognized that much important information could be obtained from such a nonmathematics specialist population.

1.5 Instrumentation

The Study entailed a large array of instruments, as indicated by Figure 1.5.1, and described below:

1. Questionnaires completed by school officials concerning school, teacher and mathematics program characteristics; organizational factors; and school and departmental policies affecting mathematics instruction.
2. Questionnaires completed by teachers to provide background information on experience, training, qualifications, beliefs and atti-tudes. Additional questionnaires to provide general information on instructional patterns (allocated time, ability of class, classroom organization and activities related to individualization of instruction, resources used; and goals and factors affecting instructional decisions), and beliefs about effective teaching. Additional ques-tionnaires related to instruction on selected specific topics.

3. Ratings by teachers (opportunity to learn) on whether the content needed to respond to each item of the achievement tests had been taught that year, in prior years, or not at all, to their students.
4. Questionnaires completed by students providing background information (e.g., parents' education and occupation), time spent on homework, and attitudes and beliefs related to mathematics.
5. Achievement tests completed by students at the end of the year.

In the Longitudinal Study, two significant additions were made to this set:

6. The achievement test and a subset of the attitudinal items were administered to the classes at the beginning of the school year.
7. Teachers of the sampled classes responded to detailed "classroom process questionnaires" dealing with how they handled subject matter during the year.

1.6   SIMS Management

IEA Studies are cooperative ventures in cross-national, or cross-system, research. The cooperation begins with the decisions of the individual systems to undertake the study. Then an international committee is formed to develop the major purposes of the study, its central questions, its design, and the various tests and questionnaires. At the same time, each participating system creates its own committee. The international and national committees interact until a common understanding is reached and the central questions, instruments and procedures are determined. Each system is free to add national options to, or omit components of, the central core of the study, so that it may address specific issues of national concern. In some cases these may be questionnaires or tests pertinent to a particular system's needs. In others, direct observations in the classroom may be included to supplement the data provided by the tests. Each system abides by the common decisions, however, just as each system agrees both to a timetable and to internationally agreed-procedures for sampling students, teachers and schools. The common instruments and approved sampling procedures exist so that comparisons across systems can be in a valid manner.

All systems cooperate in an international report and each system normally undertakes a national report. It is the series of national reports, in which each system analyses and interprets its own strengths and weaknesses in an international context, that often provides the

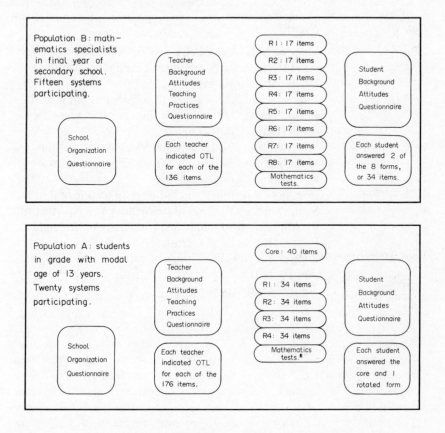

FIG 1.5.1   Survey Instrumentation for the Cross-sectional Studies in
Populations A and B

greatest value of an IEA study to each participating system. In the past, such reports have led to curriculum changes, reforms of examination systems, or new directions in teacher education.

SIMS, like all IEA Studies, required the cooperation and commitment of many institutions and individuals around the world. National Centers were asked to form National Committees to advise on the development and administration of the Study at the national level and to ensure that the responses to the Working Papers and questionnaires represented an accurate and comprehensive viewpoint for that system. Members of the National Committees included mathematics educators, teachers of mathematics, school inspectors and supervisors of mathematics, university mathematicians and educational researchers.

The following is a brief chronology of the Study (see also Appendix II).

Date	Event
1974	Meeting of IEA Council considers undertaking a Second International Mathematics Study (St. Andrews University, Scotland)
1976 (May)	International meeting held to consider major goals of a second study, (University of Illinois at Urbana-Champaign, United States)
1976 (June)	First meeting of International Mathematics Committee to outline details of Study (St. Andrews, Scotland)
1976–1977	International grid developed
1978–1979	Item pool developed (including pilot testing of items)
1980	Symposium on findings of curriculum analysis (Institute for Mathematical Didactics, University of Bielefeld, Federal Republic of Germany)
1980–1982	Data collection (For details of national schedules, see Appendix II)
1983–1987	Analysis and reporting

2

The International Grid and Item Pool

2.1 A Framework for Describing the Curriculum

In an international study of achievement, a concern commonly raised is that of the appropriateness or fairness of the tests for the students in the participating systems. This issue was referred to by Husén (1967, Vol. I, p. 21) in connection with the First International Mathematics Study:

With some justification, one might paradoxically say that the tests devised for the IEA Study are equally appropriate or inappropriate to all the systems participating in the Study.

In common with earlier IEA Studies, the conceptualization of SIMS provides a means for determining the appropriateness, or fairness, of the tests that were used. This determination can be made at the level of the intended curriculum, i.e. to what extent is the mathematics content of the test (the SIMS pool of items) "intended to be covered" by the curriculum of the target population in each system? It can also be made at the level of the implemented or taught curriculum—to what extent has the mathematical content of the test actually been taught to the students who were tested? Finally, at the student level, the amount of mathematics learned can be viewed in the light of the content of the intended curriculum and what was taught. In this chapter, the terminology used in this "curriculum analysis" is developed. The discussion is based upon the concept of the "three-tiered curriculum" as outlined in Chapter 1.

One of the earliest, and major tasks to be undertaken by the International Mathematics Committee was the development of an item pool to be used for both the construction of the international tests and the description of national curricula. On the one hand the item pool had to reflect as closely as possible the intended curriculum of each participating system. On the other hand it had to encompass a sufficient commonality of mathematical content so as to make possible meaningful between N system comparisons of what mathematics has been taught and learned. This two-fold need meant that before item development could begin, a great deal of information about national curricula had to be gathered and organized.

A key organizing framework for this work was a content-by-behavior grid. This grid helped to define the ways in which curricula were viewed in

15

each system, and across systems. Indeed, the place of this grid is so central to the Study that it can reasonably be called the bedrock upon which the entire project was built. Thus, *when mathematical content*, such as Algebra or Geometry, *is discussed in the Study it is always in terms of how this content is defined in the grid*. And when cognitive levels are referred to, they are those specified by the grid.

The grid format adopted was similar in general outline to the one used in FIMS, a two dimensional content-x-behavior array (see Table 2.4.1). The horizontal dimension (the columns) indicates the level of behavior that students were expected to be able to manifest in each of the content areas. This, in turn, provides some measure of the depth and level of understanding that was expected of the students in the target populations for that topic. For SIMS, the horizontal dimension was divided into four levels: computation, comprehension, application and analysis (Wilson, 1971: 660–662). The vertical dimension (the rows) covers the content areas. (These are discussed in greater detail in the following section.)

One of the early tasks of SIMS was a study of the mathematics curricula of several systems. The goal was to draw up a "menu" of the mathematics taught for each of the target populations that was sufficiently comprehensive that any curriculum could be considered as a selection of topics from this general collection. It was clearly understood that this was in no way to be regarded as an "ideal" curriculum but rather the union of all conceivable curricula, designed to include all topics likely to be taught in any participating system. In addition, certain topics, such as computer science, probability and statistics, were included in an attempt to anticipate possible future trends in curriculum.

The Population A content outline originally contained 133 categories under five broad classifications: Arithmetic; Algebra; Geometry; Probability and Statistics; and Measurement. As the project developed, it became apparent that Probability was not of sufficient importance internationally to warrant the label "Probability and Statistics." (Only one item on Probability was placed in the international pool.) Subsequently, this category was renamed "Statistics." Measurement was specifically identified because a number of the prospective participating systems had just undergone or were expecting to undergo conversion to the metric system. It was felt that this would provide a valuable opportunity to study whether this conversion process would manifest itself in the students' understanding of this content domain.

The content description for Population B was a similarly comprehensive outline of mathematics with 150 categories classified under nine broad headings: Sets, Relations and Functions; Number Systems; Algebra; Geometry; Analysis; Probability and Statistics; Finite Mathematics; Computer Science; and Logic. Again the outline attempted to be inclusive to make allowances for the variety of mathematics courses across schools and systems in the preuniversity year.*

Following the development of the grid, each participating system was asked to provide the International Mathematics Committee with a description of their curricula using the format of the international grid. From these descriptions an international table of specifications was compiled for each population. The development of this table required a great deal of interaction of the National Mathematics Committee with each system taking part in the Study. Unfortunately, the amount of interaction was uneven for a number of reasons. At the outset, only a small number of systems were committed to participation —essentially a subset of those in FIMS plus New Zealand, Canada and Belgium (Flemish). Moreover, communication by mail, with English as the working language, made interaction difficult, cumbersome and slow.

A major task in using the grid was to ensure, as far as possible, that in providing the requested information, different national committees understood the requests *in the same way*. Six *Working Papers* were prepared to assist in preparing responses to such requests.

Working Paper I asked National Centers for details of the various mathematics courses available to students, the proportion of students following each course, official statements on the syllabus for each course, lists of textbooks used and, where possible, copies of the textbooks. In addition, typical examination papers and standardized tests used at the target population levels were requested. Included in *Working Paper I* was a detailed description of the behavioral levels, illustrated with items from FIMS. In particular, the assumptions implicit in the use of such a content-by-behavior grid were carefully stated. The intent of this description was to provide explicit definitions of the classifications used. While not everyone would agree with the terms used to describe the levels or the number of levels, the Working Paper describes what meanings these terms would have in SIMS and National Centers were asked to respond in that spirit.

As noted earlier, the content list was divided into categories, five for Population A and nine for Population B and each of these categories was further subdivided to yield 133 topics for Population A and 150 for Population B. Each of the 133 subcategories was further subdivided whenever there was some ambiguity as to just what the International Mathematics Committee had in mind by a particular phrase or specific topic classification. Moreover, examples were included to make even more explicit what was intended by a particular term or expression.

For each cell in the grids, National Centers were asked to report on:

---

* A note should be made here concerning terminology. Although "Analysis" is used in many systems as a descriptor of that branch of mathematics dealing with functions and calculus, the same term is also used to describe one of the behavioral levels in the grid. In order to avoid confusion, "Analysis" was subsequently changed to "Elementary Functions and Calculus." Concurrently, "Sets, Relations and Functions" was renamed "Sets and Relations."

1. Its *universality* (i.e., whether the mathematical skills and knowledge defined by the cell was part of the curriculum for "all," "some" or "none" of the students in the system).
2. The *emphasis* given this aspect of the curriculum.
3. The *importance* accorded this aspect of the curriculum within each systems curriculum.

As a check on systems' understanding of the content–×–behavior grids, National Centers were also requested to supply items which they considered appropriate for ten preselected cells in each grid.

At its meeting in May 1977 the International Mathematics Committee had available the completed returns from 14 systems: Australia; Belgium (French and Flemish); Canada (Ontario); France; Hungary; Ireland; Israel; Japan; The Netherlands; New Zealand; Scotland; Sweden and the United States. The data from these systems provided the basis for the construction of the final international grids and table of specifications for the item pools. The content dimension of the grids, importance ratings for the cells and number of items allocated to each cell were based on the following criteria:

1. Overall universality of the topic.
2. Overall degree of importance accorded the topic.
3. Overall degree of importance accorded the behavior related to the topic.
4. Potential importance likely to be accorded a content area or a behavioral level as a result of impending curriculum changes.
5. Relevance of an element to issues being addressed by the Study.

On the basis of these criteria, it was decided that both computer science and logic as separate topics should be deleted from the Population B content list.

After this meeting, copies of the proposed international grids were sent to National Centers with the recommendation that the judgments of the International Mathematics Committee be considered by National Mathematics Committees and reassessed against their own curricula. It was also pointed out that the grid could be altered if systems that were still to return *Working Paper I* questionnaires had curricula differing significantly from the content areas covered by the proposed grid.

The responses to this request suggested that the proportion of content that was common to all systems was high, but both wording or placement of some elements in the content outline caused some National Committees to express doubts as to the validity of aspects of the proposed grid for their curricula. This was especially so for the geometry section where newer approaches to geometry at this level ranged from rudimentary and intuitive notions of size and space through transformations (both informal and formal) to rather abstract axiomatic treatments.

Further work developing specifications for the achievement tests took place at meetings of the IMC in 1978. At this meeting the proposed international grids were again modified in response to additional information received from National Centers, including some systems that had recently joined the Study. In addition the grids were condensed by eliminating those subtopics that were not part of the curriculum in most systems. Cells were designated "Very Important," "Important," or "Important for Some" from the within-system ratings allocated by National Centers. Later further modifications of this initial grid were agreed to in order that the grid might take into account the curricula of systems which had been strongly influenced by the Bourbaki Group.* As a result, items were introduced into the SIMS item pool that dealt with negative integer exponents and on more formal approaches to transformational geometry at the Population A level; for Population B, new topics included vector spaces and transformations in the complex plane. Although there was later to be some disagreement about the placement of specific items within behavioral categories, most national committees reported satisfaction with the eventual classification.

2.2 Behavioral Levels

These levels were used to provide some indication of the degree of understanding of a particular topic that was to be achieved by students. For example, a topic might be introduced just at that year and the student expected only to recall certain facts or carry out certain procedures. On the other hand, another topic may have been introduced in prior years and reinforced in the current year, so that students are expected to be able to solve nonroutine problems in this area. In this case, both topics would be covered and perhaps even be considered of equal importance but the expected level of behavior would be quite different for the latter topic.

The behavioral levels used in the grid were as follows†:

1. *Computation*
   Examples:
   Ability to recall specific facts
   Ability to use mathematical terminology
   Ability to carry out algorithms

---

* *Bourbaki* is the name adopted by a group of French mathematicians who set out to rewrite mathematics on an axiomatic basis. At the school level, the influence of Bourbaki is seen in a rigorous, formal treatment of mathematical concepts, with an emphasis on mathematical, especially, algebraic, structures.

† The behavioral classifications used in the study posed problems for Belgium (Flemish), where an alternative taxonomy was preferred. This added to the difficulties that their National Committee was already experiencing in relating to the listing of study content to its conception of mathematics as a field of inquiry. For others (notably, the Dutch Committee), the opinion was strongly held that reliable distinctions between the behavioral levels could not be made.

2. *Comprehension*
   Examples:
   Ability to recognize concepts
   Ability to recognize mathematical principles, rules and generalizations
   Ability to transform problem elements from one mode to another
   Ability to follow a line of reasoning
   Ability to read and interpret a problem

3. *Application*
   Examples:
   Ability to solve routine problems
   Ability to make comparisons
   Ability to analyze data
   Ability to recognize patterns, isomorphisms, and symmetries

4. *Analysis*
   Examples:
   Ability to solve nonroutine problems
   Ability to discover relationships
   Ability to construct proofs
   Ability to criticize proofs
   Ability to formulate and validate generalizations

The behavioral levels refer to the cognitive behavior of a student in carrying out a task. This in turn depends upon the background of the student as well as the focus of the task. Thus, the behavioral level of a task differs from its difficulty. For example, the exercise

$$243789265$$
$$\times 482793419$$

might be considered a difficult task but it requires only the ability to carry out a standard algorithm. Hence this exercise would likely be classified as requiring performance at the computation level of behavior.

But consider:

$$480$$
$$\times 9$$

which would be attacked simply by applying the algorithm, making the exercise one at the computation level. Alternatively, one could, recognizing that 9 is one less than 10, first find the product 4800 (as 480 × 10), then find the difference $4800 - 480$ by first subtracting 500 then adding 20, giving the result 4320. In this case, the problems would be judged to be at the application level.

On the other hand, consider this SIMS item (item 069—the code is the standard international three-digit identification code).

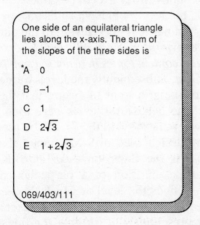

The item is not difficult as one need only add a number and its negative to zero. It is, however, first necessary to recognize that the slope of the base is zero and the slope of one of the other two sides is the negative of the other. This analysis of the data is judged to require a behavior at the application level.

Now we have item 086:

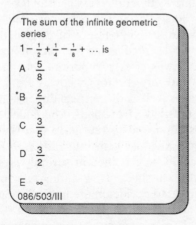

To respond to this item by the recall and use of the formula for the sum of a geometric series would be behavior at the computation level. On the other hand, if a student did not know this formula but observed that, say, on the number line, starting at 1, the series involve "jumps" halfway to 0 then halfway back to 1 then halfway back to 1/2, etc., the successive partial sums are 1/2, 3/4, 5/8, etc. Therefore, the sum of the series is greater than all distractors except B. This is behavior at the application level since it involves the ability to recognize patterns and analyze data.

It is clear from the above that it is impossible without a great deal of probing to determine the level of behavior of someone performing a task. However, this fact is not particularly relevant to the purposes of this aspect of the Study: *the behavioral levels were used to describe the goals and objectives for particular content topics in order to establish the "topography" of the International Grid.* Subsequently the levels were used to assign items to the various cells of the grid so as to ensure that the tests reflected this topography. Finally, the behavioral levels were used to classify items in order to determine where in the grid they could be used.

The item sampling design used in SIMS meant that for most systems the unit of analysis was the classroom—*not individual students.* Therefore, the behavioral classification of a particular content topic in the intended curriculum defines the level at which the *students in a class* are expected to be able to function. Thus, with respect to item 086, if the curriculum calls for presenting the formula for the sum of a geometric series then one would expect students in a class to solve this item at the computational level of behavior. That, in fact, a student may not remember a formula and have to resort to a higher level of behavior to solve the problem does not change the goal of the curriculum—it merely indicates a failure of that student to meet that "goal."

Similarly, at the system level, National Centers were to decide on the goal intended, fully recognizing that not every class in the system would achieve that goal. Some teachers may, in the press for time, omit topics; for the purposes of creating a national topography of the grid, that was not relevant.

The International Grid served as the foundation for the development of the cognitive items. It was clearly important that the item pool include items that reflected not only the topics covered but also the behavioral levels—the horizontal as well as the vertical dimension. In order to ensure proper coverage of the grid, items had to be collected, constructed and classified. Thus, it is important to recognize that the focus of classifying items was not on the individual items but rather on the cells of the grid. Thus certain items could be classified in several ways, depending on the point of view of the classifier. It was the task of the International Mathematics Committee to decide on the final classification.

The construction and selection of items was not difficult for the lower levels of cognitive behavior—computation and comprehension. The difficulties were presented at the higher levels. The multiple-choice format necessitated by the scale of the study presented some opportunities as well as challenges.

Consider item 021:

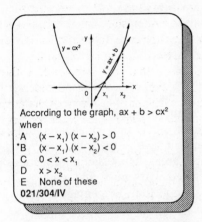

According to the graph, $ax + b > cx^2$
when
A $(x - x_1)(x - x_2) > 0$
*B $(x - x_1)(x - x_2) < 0$
C $0 < x < x_1$
D $x > x_2$
E None of these
021/304/IV

As an open-ended problem this item would involve reading the graph, a comprehension behavior. If $x_1 \times x_2$ were included as one of the choices, the item would involve merely the computation or comprehension level of behavior. However, given the choices included to solve this problem one has to analyze them and discover the relationship they determine. This involves behavior on the analysis level. In this case, the graphs of the functions become one way of communicating certain information that must be matched to other descriptions of the same data. This item provides an example of an imaginative use of the multiple-choice format in order to yield a problem at a higher level of behavior than if the open format were used.

Construction of a proof is generally an analysis level behavior and is certainly something that should be tested. However, the IMC, in consultation with many mathematicians and mathematics educators, tried a variety of multiple-choice items to test the ability to construct proofs. None was satisfactory. Hence, only a few items testing the ability to analyze a proof were included.

2.3 The Content List

As has already been indicated, the content dimension was divided into five categories and 133 subcategories for Population A and nine and 150 subcategories for Population B. In Population A, for example, one of these content areas was Arithmetic, which was further divided into nine topics:

001 Natural Numbers and Whole Numbers
002 Common Fractions
003 Decimal Fractions
004 Ratio, Proportion, Percentage

005 Number Theory
006 Powers and Exponents
007 Other Numeration Systems
008 Square Roots
009 Dimensional Analysis

These subdivisions were designated for convenience in organizing the content, with no intention of implying that each of these has, or should have, equal weight in any curriculum. Nor was this intended to reflect a division of this content category into equal parts. Indeed, each of these was further subdivided and in several cases these subdivisions broken up still further. Thus the topic areas "002 Common Fractions" and "003 Decimal Fractions" were divided into nine and eight subtopics, respectively, whereas "004 Ratio, Proportion, Percentage" was divided into two subtopics and "009 Dimensional Analysis" into three subtopics.

In the case of natural numbers and whole numbers, six subtopics were used:

001.1 Reading and writing large numbers
001.2 Operations with whole numbers
001.3 Representing whole numbers on a number line
001.4 Properties of operations
001.5 Ordering whole numbers
001.6 Estimating

Finally, these topics were also explained more fully, further subdividing when it was felt necessary. Examples were provided to further clarify the intended meaning where needed. The description of the content for Population B was not nearly so detailed. It was believed that the meanings of the terms used to list the topic were more universal and hence there were fewer dangers of ambiguities.

The content dimension was intended to provide a list of topics including *any* topic that might be covered in the curriculum of *any* participating system. The breakdown into more explicit, more precise subtopics was done to ensure comparability of the responses. Thus, while someone might take exception to the inclusion of Integers as a subtopic of Algebra (rather than Arithmetic) they could nevertheless indicate the extent to which aspects of Integers were covered in the curriculum. There was no intention of ascribing particular significance to the inclusion of Integers in Algebra rather than Arithmetic.

This content list proved, for the most part, useful in spite of some disagreement as to the placement of certain topics. Some difficulties, however, did arise. The description of the category Geometry proved to be inadequate to cope with the great variety of topics and approaches to the teaching of geometry in the participating systems. Indeed, some systems found that the content list for Geometry implied an approach to teaching the topic that was quite alien, and one with which they were not in sympathy. The discovery of so much diversity is an important fact in itself. In practice, it created some major difficulties for a number of systems with respect to describing their curricula.

Another difficulty centered on the structure of the content list. The organization of topics on the list did not correspond to that of some systems. For example, in the content list, common fractions appeared before decimal fractions. In some systems, the whole numbers or counting numbers, 1,2,3 . . . are first introduced, then extended to the natural numbers, that is, zero is added. Common fractions, of the form a/b where a is a natural number and b is a whole number (that is, different from zero) are treated before decimal fractions (repeating and nonrepeating infinite decimals) are introduced.

In other systems, finite decimal fractions are introduced after the natural numbers and then extended to infinite decimals. Indeed the real numbers are defined as decimal numbers, finite and infinite. Rational numbers are then presented as a subset of the real numbers. The order of this development is at odds with the order in which the topics were listed in the content dimension of the grid. An order consistent with both these developments would clearly have been impossible.

2.4 The International Grid: Population A

As already noted, each National Center was requested to provide "importance ratings" of each behavioral level for each content listing. An international consensus on importance was determined from the individual responses by the International Mathematics Committee, and the cognitive item pool was then developed from this international grid; the importance ratings served as weights to help determine the number of items required for each code in the international grid. What follows is a description of the content categories along with sample items to illustrate the topics and behavioral levels.

000 Arithmetic

The Arithmetic content area is important for Population A because students at this level have studied arithmetic for several years. Arithmetic includes topics that most children have had an opportunity to master. Topics considered most important internationally include Whole Numbers (001), Common Fractions (002), Decimal Fractions (003), and Ratio, Proportion and Percentage (004).

Which of the following operations with whole numbers will ALWAYS give a whole number?

I   Addition
II Multiplication
III Division

A I only   * D I and II only
B II only   E II and III only
C III only

137/001/II

$\frac{3}{8} - \frac{1}{5}$ is equal to

A $\frac{1}{20}$ D $\frac{19}{40}$

* B $\frac{7}{40}$ E $\frac{2}{3}$

C $\frac{7}{20}$

187/002/I

A car takes 15 minutes to travel 10 kilometers. What is the speed of the car?

A 30 kilometers per hour
* B 40 kilometers per hour
C 60 kilometers per hour
D 90 kilometers per hour
E 150 kilometers per hour

192/004/III

100 Algebra

The Algebra content area has the potential for significant international variation across topics. Some students are engaged in the study of algebra topics during the Population A year, while others do not begin a study of algebra until the following year. The Algebra topics considered internationally important at the Population A level included Integers (101), Rational Numbers (102), Algebraic Expressions (104), Linear Equations/Inequations (106), and Functions (107).

Illustrative Algebra items are:

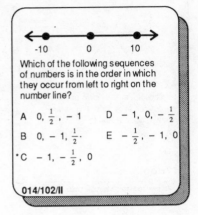

Which of the following sequences of numbers is in the order in which they occur from left to right on the number line?

A $0, \frac{1}{2}, -1$

B $0, -1, \frac{1}{2}$.

*C $-1, -\frac{1}{2}, 0$

D $-1, 0, -\frac{1}{2}$

E $-\frac{1}{2}, -1, 0$

014/102/II

The cost of printing greeting cards consists of a *fixed charge of 100 cents* and a charge of 6 cents for each card printed. Which of the following equations can be used to determine the cost of printing n cards?

*A cost = $(100 + 6n)$ cents
B cost = $(106 + n)$ cents
C cost = $(6 + 100n)$ cents
D cost = $(106n)$ cents
E cost = $(600n)$ cents

052/104/III

m	−1	1	2	4
n	−1	3	5	9

For the table shown, a formula that could relate m and n is

A $n = m$
B $n = 3m$
C $n = -m^2 + 1$
D $n = m^2 + 1$
*E $n = 2m + 1$

055/107/III

200 Geometry

Geometry showed a high potential for international variation in approaches to the topic. Only a few geometry topics were considered universally important and most topics were important in some systems and not in others. The following sample of items shows the variety of content available for study.

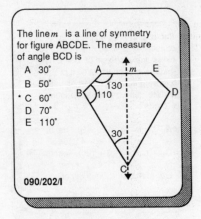

The line *m* is a line of symmetry for figure ABCDE. The measure of angle BCD is

A   30°
B   50°
* C   60°
D   70°
E   110°

090/202/I

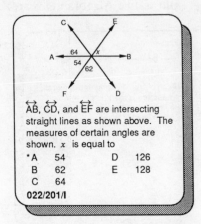

$\overleftrightarrow{AB}$, $\overleftrightarrow{CD}$, and $\overleftrightarrow{EF}$ are intersecting straight lines as shown above. The measures of certain angles are shown. *x* is equal to

*A   54   D   126
B   62   E   128
C   64

022/201/I

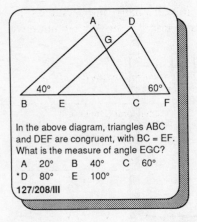

In the above diagram, triangles ABC and DEF are congruent, with BC = EF. What is the measure of angle EGC?

A   20°   B   40°   C   60°
*D   80°   E   100°

127/208/III

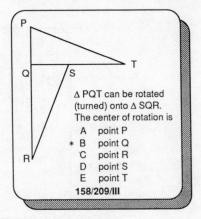

Δ PQT can be rotated (turned) onto Δ SQR. The center of rotation is

A   point P
* B   point Q
C   point R
D   point S
E   point T

158/209/III

300 Statistics

This content area was introduced into SIMS not only to reflect achievement on a recent addition to the mathematics curriculum but also for its potential to serve as a baseline for future studies where statistics might have a more prominent role. The internationally important topics are Collection (301), Organization (302), Representation (303) and Interpretation of Data (304).

The petals on 100 flowers of different
kinds were carefully counted, and the
results are shown in this table.

No. of Petals	Frequency
10-12	5
13-15	22
16-18	48
19-21	18
22-24	7

How many of the flowers had FEWER
than 19 petals?

   A 48   B 52   C 73
* D 75   E 93

131/302/II

400 Measurement

The Measurement content area is internationally important as an application of elementary arithmetic. Most Population A students have

Table 2.4.1   *Population A: Importance for Instrument Construction of Content Topics and Behavioral Categories*

Content topics	Behavioral Categories* Computation	Comprehension	Application	Analysis
000 Arithmetic				
001 Natural numbers and whole numbers	V	V	V	I
002 Common fractions	V	V	I	I
003 Decimal fractions	V	V	V	I
004 Ratio, proportion, percentage	V	V	I	I
005 Number theory	I	I	–	–
006 Powers and exponents	I	I	–	–
007 Other numeration systems	–	–	–	–
008 Square roots	I	I	–	–
009 Dimensional analysis	I	I	–	–
100 Algebra				
101 Integers	V	V	I	I
102 Rationals	I	I	I	I
103 Integer exponents	Is	–	–	–
104 Formulas and algebraic expressions	I	I	I	I
105 Polynomials and rational expressions	I	Is	–	–
106 Equations and inequations (linear only)	V	I	I	Is
107 Relations and functions	I	I	I	–
108 Systems of linear equations	–	–	–	–
109 Finite systems	–	–	–	–
110 Finite sets	I	I	I	–
111 Flowcharts and programming	–	–	–	–
112 Real numbers	–	–	–	–

Table 2.4.1 (cont'd)   *Population A: Importance for Instrument Construction of Content Topics and Behavioral Categories*

Content topics	Behavioral categories* Computation	Comprehension	Application	Analysis
200 Geometry				
201 Classification of plane figures	I	V	I	Is
202 Properties of plane figures	I	V	I	I
203 Congruence of plane figures	I	I	I	Is
204 Similarity of plane figures	I	I	I	Is
205 Geometric constructions	Is	Is	Is	–
206 Pythagorean triangles	Is	Is	Is	–
207 Coordinates	I	I	I	Is
208 Simple deductions	Is	I	I	I
209 Informal transformations in geometry	I	I	I	–
210 Relationships between lines and planes in space	–	–	–	–
211 Solids (symmetry properties)	Is	Is	Is	–
212 Spatial visualization and representation	–	Is	Is	–
213 Orientation (spatial)	–	Is	–	–
214 Decomposition of figures	–	–	–	–
215 Transformational geometry	Is	Is	Is	
300 Statistics				
301 Data collection	Is	I	I	–
302 Organization of data	I	I	I	Is
303 Representation of data	I	I	I	Is
304 Interpretation of data (mean, median, mode)	I	I	I	–
305 Combinatoric	–	–	–	–
306 Outcomes, sample spaces and events	Is	–	–	–
307 Counting of sets, P(A B), P(A B), independent events	–	–	–	–
308 Mutually exclusive events	–	–	–	–
309 Complementary events	–	–	–	–
400 Measurement				
401 Standard units of measure	V	V	V	–
402 Estimation	I	I	I	–
403 Approximation	I	I	I	–
404 Determination of measures: areas, volumes, etc	V	V	I	I

*Rating scale: V = very important; I = important; Is = important for some systems. A dash (–) = not important

had some exposure to measurement concepts. Furthermore, since some systems adopted the metric system since FIMS, some interesting comparisons are possible. All measurement topics showed some international importance: Standard Units (401), Estimation (402), Approximation (403), and Mensuration (404).

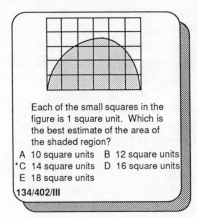

Each of the small squares in the figure is 1 square unit. Which is the best estimate of the area of the shaded region?
A 10 square units   B 12 square units
*C 14 square units   D 16 square units
E 18 square units
134/402/III

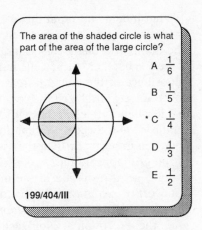

The area of the shaded circle is what part of the area of the large circle?

A $\frac{1}{6}$

B $\frac{1}{5}$

*C $\frac{1}{4}$

D $\frac{1}{3}$

E $\frac{1}{2}$

199/404/III

Table 2.4.1 presents the international grid for Population A. It should be remembered that the importance ratings represent a consensus. As such, it is to be expected that the curriculum for each system in the Study will be reflected in the grid more or less well, depending on the extent to which systems have unique emphases on certain aspects of mathematics. Since, as is discussed in Sections 2.1 and 2.2, the grid served as the blueprint for the international tests (that is, served as a guide to the number and kind of items to be included), it is important to envision the "goodness of fit" of this grid to each system's curriculum.

In order to provide some information on this "goodness of fit," consider Figure 2.4.1. The figure is designed to illustrate the match between each system's curriculum and the international grid as follows. First, each importance rating is given a numerical value:

V (very important) 3
I (important) 2
Is (important in some systems) 1
Not important or not taught 0

On the basis of these values, the difference between the importance rating and the international rating was computed for each topic area (row) in the grid, and for each system. Figure 2.4.1 presents the results of these computations. The cross-hatched regions indicate that the international rating for the given cell exceeded the rating provided by the National Mathematics Committee for the indicated system. The black regions indicate that the rating of the National Committee for that topic exceeded the rating in the international grid. The width of the shaded region, ranging from one to three units, reflects the magnitude of the difference in ratings. Blank cells indicate agreement between national and international importance ratings for the corresponding topics.

What judgments about the overall fit between the international grid and the national grids for each system emerge? Overall, the fit was rather good. The number of blank cells is large and there are many cross-hatched cells as well. Together, these cells indicate that for the majority of topics, the importance of the content areas (and, hence, the number of items in the pool) reflects fairly well the content emphasis for most systems.

The situation seems best in *Measurement*. In New Zealand, Canada (Ontario) and Sweden, estimation and approximation were rated slightly higher in importance than they were in the grid. Otherwise, the coverage of the grid appears fairly good. In *Statistics*, there appears to have been, overall, good provision in the international grid for most countries. The noteworthy exception is Scotland, where more importance is given to sets than is found in the grid. *Arithmetic* is well covered in the international grid; the exception is NonDecimal Numeration System, which is given more importance in some countries (notably, Hungary, Ireland, Japan, Swaziland and Thailand) than in the grid.

For *Algebra*, there is some indication that the international grid does not give enough importance to some topics in some systems. Exponents, Flowcharting and Programming, and Real Numbers. Countries having notable "block bars" included Finland, France, Japan, The Netherlands, New Zealand and Canada (Ontario).

The only topic for which there appears to be a substantial problem of mismatch is *Geometry*. Indeed, a major finding of the Study proved to be the great diversity of curricula in geometry for Population A around the world. In spite of the best efforts of the International Mathematics Committee to provide a comprehensive covering for geometry, there remained some systems, and some topics, for which the match was not good.

2.5   The International Grid: Population B

A parallel outline of mathematical content and behavioral levels was constructed for Population B with a great deal of effort made to produce a content list reflecting all of the mathematics that might be taught to students completing a program in secondary schools. Sample items from the cells of this grid will illustrate its construction and show how the international pool of cognitive items was developed. (See Table 2.5.1.)

Sets and Relations (100): The concept of Function (104) was judged internationally to be the most important category within this topic.

Number Systems (200): Items were written to cover the topics of Natural Numbers (202), Decimals (203), Real Numbers (204), and Complex Numbers (205). The following items illustrate how the various number systems were represented.

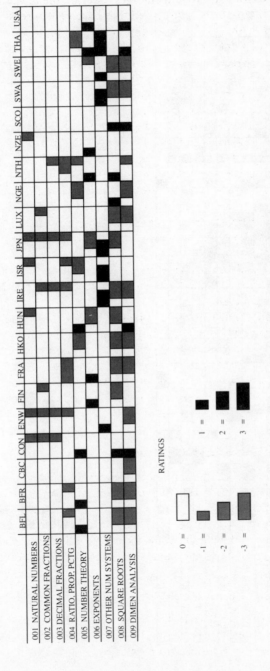

FIG 2.4.1.A The Match Between the International Grid and the National Grids: Population A (ARITHMETIC)

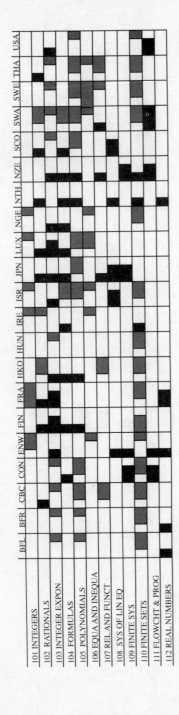

Fig 2.4.1.B   The Match Between the International Grid and the National Grids: Population A
(ALGEBRA)

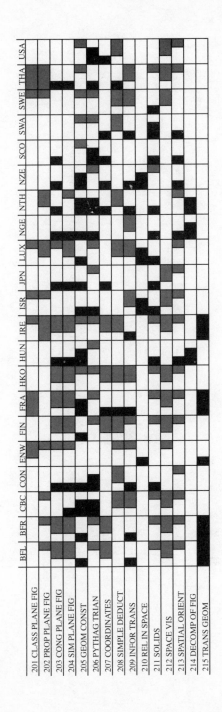

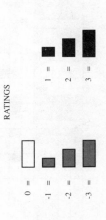

FIG 2.4.1.C   The Match Between the International Grid and the National Grids: Population A (GEOMETRY)

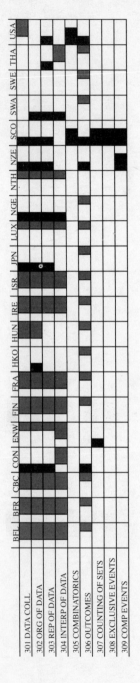

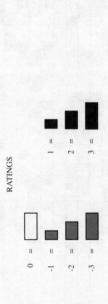

Fɪɢ 2.4.1.D   The Match Between the International Grid and the National Grids: Population A
(STATISTICS)

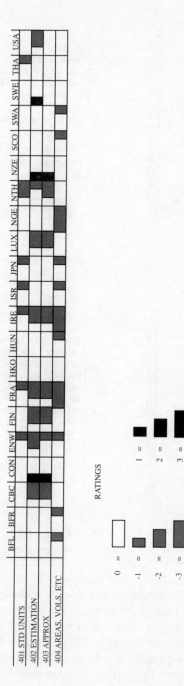

FIG 2.4.1.E   The Match Between the International Grid and the National Grids: Population A (MEASUREMENT)

A function f with domain $\{1, 2, 3\}$ is defined by $f(x) = 2^x$. The range of f is

A $\{\frac{1}{4}, \frac{1}{2}, \frac{1}{8}\}$ B $\{\frac{1}{2}, 1, 1\frac{1}{2}\}$

C $\{1, 2, 4\}$ *D $\{2, 4, 6\}$

D $\{2, 4, 8\}$

046/104/I

A real valued function f defined on a set of real numbers is said to be additive if, for all x and y in the domain of f,
$f(x + y) = f(x) + f(y)$.
If the function f defined on the set of positive real numbers by $f(x)$ is equal to
A x^2, then f is additive
B $\sin x$, then f is additive
C $\log_{10} x$, then f is additive
D $2x$, then f is additive
*E None of the above is additive

076/104/11

If a is a digit, let $.\bar{a}$ represent the number with decimal expansion $.aaaaaa\ldots$.
What is
$.\bar{7} + .\bar{4}$?
A 1.1   B $1.\bar{1}$
C 1.2   *D $1.\bar{2}$
E 1.3

094/203/IV

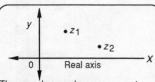

The complex number $z = x + iy$ (where x and y are real numbers) can be represented by the point (x, y). In the diagram, z_1 and z_2 represent two complex numbers. If z_3 is selected such that the origin, z_1, z_3 and z_2 are the consecutive vertices of a parallelogram, then z_3 is equal to

A $z_1 z_2$ B $\dfrac{z_1}{z_2}$

*C $z_1 + z_2$ D $z_1 - z_2$

E $\sqrt{z_1^2 + z_2^2}$

034/205/IV

Algebra (300): Algebra is important because its techniques and concepts are developed throughout the years between Population A and Population B. The topics in algebra that were considered most important include Polynomials (301), Roots and Radicals (303), Equations and Inequalities (304) and Systems of Equations (305). Algebraic Representation, Problem Translation, Graphing, and Properties of Algebraic Expressions appear within the item pool.

A stationer wants to make a card
8 cm long and of such a width that
when the card is cut into halves,
the original width becmes the length
and the shape of each half is similar
to the original card. What width, in
centimeters, should he make the
original card?

A   4   *B $4\sqrt{2}$ C $5\sqrt{2}$

D $5\sqrt{3}$ E   6

007/304/IV

Geometry (400): As in Population A, the content of geometry showed the
most international variety. Analytic (coordinate) Geometry (403) and
Trigonometry (406) are commonly included, but systems differ widely on
the use of Synthetic Euclidean Geometry (401), Vector Geometry (405),
and Transformational Geometry (409). Items were developed to include
all of these aspects of geometry.

The rectangular coordinates of three
points in a plane are Q (–3, –1),
R (–2, 3) and S (1, –3). A fourth
point T is chosen so that $\overrightarrow{ST} = 2\overrightarrow{QR}$.
The y- coordinate of T is

A   –11   B   –7   C   –1

D   1   *E   5

022/405/II

Let l and m be two intersecting lines
in the Euclidean plane, and let $\vec{v}$ be a
non-zero vector. Let S_l indicate a
reflection in the line l , S_m indicate a
reflection in the line m , and T indicate a translation of vector $\vec{v}$.

Which of the following transformations
is a rotation?

A $S_l \circ T$ *B $S_m \circ S_l \circ T$

C $S_l \circ T \circ S_m \circ S_l \circ T$

D $T \circ S_m$ E $S_m \circ S_l \circ S_m$

130/401/I

It is important to note that these items may represent more than one cell in
the grid. In this respect, the grid serves to help insure that topics
considered internationally important are not unrepresented, but it does not
attempt to provide equal representation of topics. A good illustration of
this is given by the vector approach where items include ordered n-tuples,
matrix representation, and arrows in a plane. It can be argued that such
items involve coordinate systems, algebraic representation, and synthetic
methods, although their grid classification is Geometry-Vector Methods.

Find the difference $\vec{b} - \vec{a}$ of the vectors

$$\vec{a} = \begin{pmatrix} 4 \\ 2 \end{pmatrix} \text{ and } \vec{b} = \begin{pmatrix} 0 \\ 3 \end{pmatrix}$$

A $\begin{pmatrix} -4 \\ -2 \end{pmatrix}$ *B $\begin{pmatrix} -4 \\ 1 \end{pmatrix}$ C $\begin{pmatrix} 4 \\ -1 \end{pmatrix}$

D $\begin{pmatrix} 4 \\ 2 \end{pmatrix}$ E $\begin{pmatrix} 4 \\ 5 \end{pmatrix}$

052/405/I

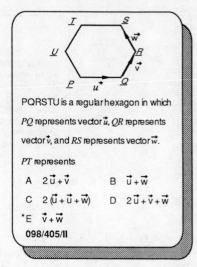

PQRSTU is a regular hexagon in which PQ represents vector $\vec{u}$, QR represents vector $\vec{v}$, and RS represents vector $\vec{w}$.

PT represents

A $2\vec{u} + \vec{v}$ B $\vec{u} + \vec{w}$

C $2(\vec{u} + \vec{u} + \vec{w})$ D $2\vec{u} + \vec{v} + \vec{w}$

*E $\vec{v} + \vec{w}$

098/405/II

Elementary Functions and Calculus (500): This content area, which includes Functions (501), Limits (503), Derivatives and their Applications (504, 505) and Integrals and their Applications (506, 507, 508) is internationally considered to be the most important topic in the grid. These topics are the culmination of most secondary programs and include the content of the Population B year common to most systems. Again, a wide variety of content is represented, including functional notation, graphical representation, and the definitions and problems of the calculus.

If $\log N = n$, then $\log N^2$ is equal to

A $n + 2$

B n^2

C $\dfrac{n}{2}$

*D $2n$

E $n - 2$

087/501/II

If $xy = 1$ and x is greater than 0, which of the following statements is true

A When x is greater than 1, y is negative.
B When x is greater than 1, y is greater than 1.
C When x is less than 1, y is less than 1.
D As x increases, y increases.
*E As x increases, y decreases.

070/502/II

The velocity of a body moving in a straight line t seconds after starting from rest is $(4t^3 - 12t^2)$ meters per second. How many seconds after starting does its acceleration become zero?

  A 1
*B 2
  C 3
  D 4
  E 6

088/505/III

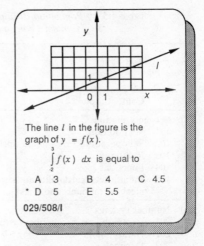

The line l in the figure is the graph of $y = f(x)$.

$\int_{-2}^{3} f(x)\ dx$ is equal to

  A 3   B 4   C 4.5
*D 5   E 5.5

029/508/I

Probability and Statistics (600): The study of Probability (601) is an important mathematical topic internationally. The topics of Statistics (602) and Statistical Distribution (603) are generally considered important and usually included as an application of mathematics.

Four persons whose names begin with different letters are placed in a row, side by side. What is the probability that they will be placed in alphabetical order from left to right?

  A $\dfrac{1}{120}$ *B $\dfrac{1}{24}$

  C $\dfrac{1}{12}$ D $\dfrac{1}{6}$

  E $\dfrac{1}{4}$

045/601/I

The mean of a population is 5 and its standard deviation is 1. If 10 is added to each element of the population, the new mean and standard deviation are

*A mean = 15
     standard deviation = 1
 B mean = 15
     standard deviation = 5
 C mean = 15
     standard deviation = 11
 D mean = 10
     standard deviation = 1
 E mean = 10
     standard deviation = 5

074/602/II

Finite Mathematics (700): Only Combinatorics (701) was judged of sufficient international importance to be included in the item pool.

Table 2.5.1   *Population B: Importance for Instrument Construction of Content Topics and Behavioral Categories*

Content topics	Behavioral categories* Computation	Comprehension	Application	Analysis
100 Sets and relations				
101 Set notation	I	I	–	–
102 Set operations (e.g., union, inclusion)	I	I	–	–
103 Relations	–	–	–	–
104 Functions	V	V	V	I
105 Infinite sets, cardinality and cardinal algebra (rationals and reals)	–	–	–	–
200 Number systems				
201 Common laws for number systems	I	I	I	–
202 Natural numbers	I	I	I	I
203 Decimals	I	I	I	I
204 Real numbers	I	I	I	–
205 Complex numbers	V	I	I	I
300 Algebra				
301 Polynomials (over R)	V	V	V	I
302 Quotients of polynomials	I	I	I	–
303 Roots and radicals	V	V	I	–
304 Equations and inequalities	V	V	V	I
305 System of equations and inequalities	V	V	V	I
306 Matrices	Is	Is	Is	Is
307 Groups, rings and fields	–	–	–	–
400 Geometry				
401 Euclidean (synthetic) geometry	I	I	–	–
402 Affine and projective geometry in the plane	–	–	–	–
403 Analytic (coordinate) geometry in the plane	I	I	V	I
404 Three-dimensional coordinate geometry	–	–	–	–
405 Vector methods	I	I	I	I
406 Trigonometry	V	V	V	I
407 Finite geometries	–	–	–	–
408 Elements of topology	–	–	–	–
409 Transformational geometry	Is	Is	Is	Is
500 Elementary functions and calculus				
501 Elementary functions	V	V	V	V
502 Properties of functions	V	V	V	I
503 Limits and continuity	I	I	I	–
504 Differentiation	V	V	I	I
505 Applications of the derivative	V	V	V	I
506 Integration	V	V	V	I
507 Techniques of integration	V	V	I	I
508 Applications of integration	V	V	V	I
509 Differential equations	Is	Is	Is	Is
510 Sequences and series of functions	–	–	–	–

Table 2.5.1 (cont'd)   *Population B: Importance for Instrument Construction of Content Topics and Behavioral Categories*

Content topics	Behavioral Categories* Computation	Comprehension	Application	Analysis
600 Probability and statistics				
601 Probability	V	V	I	–
602 Statistics	I	I	I	–
603 Distributions	I	I	I	–
604 Statistical inference	Is	Is	–	–
605 Bivariate statistics	–	–	–	–
700 Finite mathematics				
701 Combinatorics	I	I	I	–
800 Computer science	Is	Is	I	–
900 Logic	–	–	–	–

*Rating scale: V = very important; I = important; Is = important for some systems. A dash (–) = not important

"Goodness of fit" of Population B Grid to National Curricula

Figure 2.5.1 depicts the degree of fit of the Population B grid (Table 2.5.1) to the Population B curriculum for each SIMS system. Overall, the international grid appears to contain each system's curriculum rather well. For example, Finite Mathematics and Probability and Statistics are encompassed adequately by the international grid. (A noteworthy exception is for Japan where Probability and Statistics is rated higher than in the international grid.) Algebra and Elementary Functions and Calculus appear to fare well, too. In Algebra, the exception is that Matrices and Groups and Rings and Fields receive more importance in several countries than is reflected in the international grid. In Elementary Functions and Calculus, the topics of Limits and Continuity, and Sequences and Series were rated somewhat higher than appears in the international grid.

In Sets and Relations, the topic of Relations receives less attention in the grid than in several countries. In Geometry, the fit of the international grid is better than in Population A. Only in Solid Geometry and Transformational Geometry is there significant lack of fit, and this occurs for less than one-half of the countries. Overall, in view of the diversity between systems, the international grid appears to be a creditable job of reflecting the curricular content of each participating system.

2.6   The International Item Pool

In developing the SIMS item pool, the major factors taken into account were:

1. The need to sample adequately the universe prescribed by the international grids with items that were as free as possible from cultural bias;
2. The need to include sufficient FIMS of achievement items to allow appropriate comparisons to be made over the 20-year period between that study and SIMS;

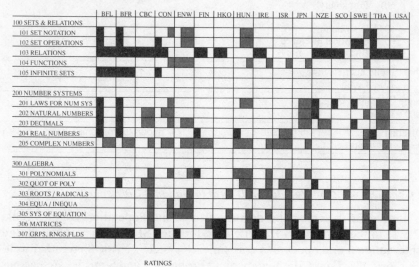

FIG 2.5.1.A
The Match Between the International Grids and the National Grids:
Population B

3. The need to oversample items from particular fields of interest in the Study. In addition to a number of content and behavior subscores, there were hypotheses which called for subscores based on items common to both the FIMS and SIS:

   items for which the use of a calculator was appropriate
   minimal competency/basic skill items
   items on estimation and approximation
   items involving proportional thinking
   "new mathematics" items included in FIMS
   items in which SI measurements were used.

For those systems participating in the longitudinal (classroom processes) phase of the Study, there was a fourth interest:

4. The need to include items on topics that were likely to be sensitive to growth during the school year. These topics were determined to be:

   Fractions (common and decimal);
   Ratio (Proportion and Percent);
   Algebra (Integers, Formulae and Equations);
   Geometry;
   Measurement.

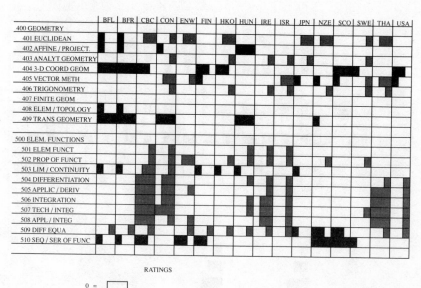

RATINGS

FIG 2.5.1.B
The Match Between the International Grids and the National Grids:
Population B

The need to have an adequate number of items on which to form subscores necessitated additional items for these topics and hence the greater weighting for arithmetic in the longitudinal version of the Population A Study than in the cross-sectional version.

Measurement Criteria

As items were selected for the international item pool, the following criteria were taken into account:

1. In general, each item should have a mean difficulty in the pilot testing phase in the range 0.40–0.90 and a range of difficulties across systems.
2. Each item should have a mean discrimination in the pilot testing phase of not less than 0.30 and discriminate well in almost all systems in which it was pilot tested.
3. The items that were rated by National Centers as likely to be sensitive to growth should be included. These were of particular interest to systems taking part in the longitudinal phase of the Study.

	BFL	BFR	CBC	CON	ENW	FIN	HKO	HUN	IRE	ISR	JPN	NZE	SCO	SWE	THA	USA
600 PROB. & STATISTICS																
601 PROBABILITY																
602 STATISTICS																
603 DISTRIBUTIONS																
604 STAT INFERENCE																
605 BIVAR STAT																
700 FINITE MATHEMATICS																
701 COMBINATORICS																
800 COMPUTER SCIENCE																
900 LOGIC																

RATINGS

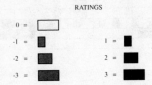

0 =	☐		
-1 =	■	1 =	■
-2 =	■	2 =	■
-3 =	■	3 =	■

Fig 2.5.1.C
The Match Between the International Grids and the National Grids:
Population B

Care was taken in the selection of items to try to minimize the likelihood of "floor" and "ceiling" effects on subtests. The diversity of curricula and the variation in ordering of topics in curricula made this a difficult task. Different systems may be thereby affected from subtest to subtest.

Item Distribution

Tables 2.6.1 and 2.6.2 below show the numbers of items by behavior in each of the broad content areas.*

* The classifications are those reflected in the cognitive item tables that are used throughout this volume. (See Volume 2.)

Issues in Item Development

Item Classification

As part of the item development phase, National Committees gave their judgments as to which content and which behavior each item was designed to measure. For the most part, these opinions confirmed the judgment of the

Table 2.6.1 *(a) Population A—Cross-sectional Study: SIMS Item Pool*

Content	Behavior				
	Computation	Comprehension	Application	Analysis	Totals
Arithmetic	17	15	11	3	46
Algebra	18	10	8	4	40
Geometry	12	16	18	2	48
Statistics	5	8	4	1	18
Measurement	7	7	9	1	24
Totals	59	56	50	11	176

(b) Population A—Longitudinal Study: SIMS Item Pool

Content	Behavior				
	Computation	Comprehension	Application	Analysis	Totals
Arithmetic	24	18	18	2	62
Algebra	14	11	6	1	32
Geometry	10	12	17	3	42
Statistics	5	10	2	1	18
Measurement	7	7	11	1	26
Totals	60	58	54	8	180

Notes: 1. The total number of items in the Population A pool, for either the Cross-sectional or Longitudinal Study (or both) was 199.
2. The number of items common to both versions of the Study was 157.

Table 2.6.2 *Population B: SIMS Item Pool*

Content	Behavior				
	Computation	Comprehension	Application	Analysis	Totals
Sets, Relations and Functions	2	3	1	1	7
Number Systems	5	5	6	3	19
Algebra	9	6	7	3	25
Geometry	9	8	8	3	28
Elementary Functions/Calculus	16	14	13	3	46
Probability and Statistics	4	10	3	–	17
Finite Mathematics	–	2	2	–	4
Totals	45	48	40	13	146

International Mathematics Committee with respect to the content being measured. The small number of items on which there were between-system differences derived from variations in traditions of classification of mathematical content. Thus, a Population B item classified under Elementary Functions in one group of systems might be regarded as belonging in Coordinate Geometry in another group of systems. In some of these cases, there was disagreement, not so much with the subtopic covered but with the general classification to which that subtopic should be assigned. Nevertheless, whatever the opinions of the National Committees on the differences between the International Grid and a grid which they would construct for their own system, there was fairly general agreement on classification of items with respect to content.

The classification of items by behavior produced divergence of opinion between systems on a significant number of items. The appropriate behavioral classification of an item for an individual student or class obviously depends on prior learning, the curriculum to which the student has been exposed and emphases within that curriculum. Assumptions have to be made, for a target population within a system, about the background most students bring to a task and how they will respond to the task. Thus, an item involving manipulation with common fractions might be judged to be at the computation level in a system (such as the United States) where the skill is taught in the Population A year and at the analysis level in systems (such as France) where major emphasis is given in Population A to decimal fractions and the manipulation of common fractions is not taught until later than the Population A year. But in spite of these potential problems there was a remarkable degree of agreement on the classification of computation items and analysis items. Differences in preferred classification, where they occur, were mainly with items classified as belonging to the categories of comprehension and application.

Translation Difficulties

During item development, National Centers drew attention to items which presented translation difficulties. Problems highlighted included:

1. Occasional examples in which two different English terms (e.g., symmetry/reflection, number/frequency) translated to a single term in other languages. Thus, for example, the French translate "line of symmetry" by "axe de symetrie."

However, "reflection" as in 096

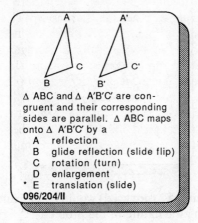

$\triangle$ ABC and $\triangle$ A'B'C' are congruent and their corresponding sides are parallel. $\triangle$ ABC maps onto $\triangle$ A'B'C' by a

A   reflection
B   glide reflection (slide flip)
C   rotation (turn)
D   enlargement
* E   translation (slide)

096/204/II

(distractor A) is translated as "symetrie axiale" whereas "reflection" in the stem

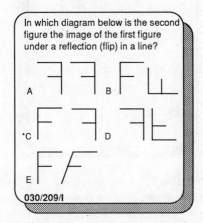

In which diagram below is the second figure the image of the first figure under a reflection (flip) in a line?

A

B

*C

D

E

030/209/I

is translated as "symetriques par rapport a un droite."

Through the use of back translation every attempt was made to provide translations that preserved the original intent of the items.

2. Inability to translate items involving amounts of money where a decimal point is used (e.g., $1.38) to an equivalent in systems such as Japan where no decimal point is used in everyday money transactions. Item T1 (not used) offered a useful way of presenting a problem of adding decimals. An alternative could be found by using measurement situations.

```
Add:            $3.06
                10.00
                 9.14
                 5.10
The total is
   A            $2730
   B            $17.20
   C            $17.30
   D            $27.20
 * E            $27.30

T1/003/I
```

On the other hand, T2 (not used) would be difficult to translate to yen without changing the nature of the problem.

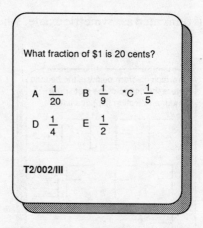

While the problem in T3 (not used) of translating this into other currencies is not too great, there was a cultural question as to whether children in certain systems would be familiar enough with parking lots to handle this question.

A parking lot charges 35 cents for
the first hour and 25 cents for each
additional hour or fraction of an hour.
For a car parked from 10:45 in the
morning until 3:05 in the afternoon,
how much money should be charged?

A $1.10   B $1.25
*C $1.35   D $1.60
E $1.75

T3/003/III

Each of the above items were rejected. However, item 190 was
included in the international item pool.

Cloth is sold by the square meter. If
6 square meters of cloth cost $4.80,
the cost of 16 square meters will be

*A $12.80 B $14.40
C $28.80 D $52.80
E $128.00

190/004/III

3. Inability to translate some of the ideas expressed in item stems into
   Chinese (Cantonese) in such a way that the same meaning would be
   conveyed to students.

Most items presenting these sorts of difficulty were rejected, but in a few
cases (e.g., some problems involving money) the items were retained since
they formed part of important subtests.

Cultural Bias

National Centers were instructed to identify items for which cultural
considerations would be likely to make them more difficult in their systems
than in others. This resulted in the rejection of a number of items. Most

items affected were those concerned with the utilization of mathematical knowledge and skills in "everyday" situations. Finding "everyday" situations which would have meaning for students from each of a wide range of cultures proved quite difficult. It was pointed out that in some systems students would be unlikely to be able to relate to the notion of a fence around a garden. Some items testing knowledge of profit percent were discarded or rewritten because the implied profit margins were too high to be acceptable in certain cultures.

Interestingly, some items that might have been expected to have disadvantaged students in some systems were not commented on. This is probably due to the fact that the textbooks in use in those systems contain examples of the type employed in the SIMS tests. But it must be conceded that for these students some items refer to a vicarious sort of "everyday" situation. During translation, National Centers were able to change such elements as proper names, place names and monetary units to those with which their students would be familiar where this seemed helpful and where the nature and relative difficulty of the item were judged to have been unaffected.

Effects of Tracking in the Development of the Grid

Different courses, or "tracks," are found in some systems. At the Population A level, where different courses exist (such as in The Netherlands where students may be in any of five school types each with its own curriculum) National Committees were faced with a dilemma in responding to Working Papers and questionnaires—whether to try to make blanket responses or whether to make a separate response for each school type. Typically, the responses were given in considerable detail with separate data for each school type. But for the construction of the international grids, the International Mathematics Committee had to make decisions about the relative importance of topics, based on estimates of the proportion of students having the opportunity-to-learn the topic in such systems. Results from pilot testing seemed to support these judgments. In The Netherlands, item statistics from a carefully drawn sample from all school types were confirmatory and were similar to those that could reasonably be expected from an ostensibly uniform system making the same importance ratings for cells of the international grid.

At the Population B level, most systems have more than one mathematics course available to students, and in some cases students may take more than one of these courses. Again, the systems supplied considerable detail on the courses available and on the proportion and quality of students taking them. Decisions on importance ratings for the Population B international grid were assisted by this information.

The most difficult decisions were for those topics such as Statistics and Probability which in several systems are taught in depth to a relatively small proportion of the population and in some systems are not taught at all. The "Is" rating (Important for Some) was applied to cells of the grid in such cases.

2.7 Summary

In SIMS the content-by-behavior grid provided a key organizing framework for surveying the mathematics curriculum for each target population in each system. On the basis of detailed information requested as the grid was developed, consensus ratings were arrived at for the importance of each cell in the grid for each of the two target populations. As a result, international grids were devised. These grids, in turn, served as blueprints for the construction of the international achievement tests. The importance ratings (internationally-determined consensus ratings) were regarded as weights for determining the number of items on the tests that were required for each cell in each of the two grids. The result of the international grid and test development exercise was a pool of 199 items at the Population A level and 136 items at the Population B level.

3

National Characteristics of Educational Systems

Chapter 2 has described the development of the item pool that provided the basic starting point for all aspects of SIMS. In this chapter the Study's other starting point, the populations that were investigated, are described.

As seen in Chapter 1, SIMS was targeted on two populations defined for planning purposes as:

Population A: All students in the grade in which the modal number of students has attained the age of 13.0–13.11 years by the middle of the school year.

Population B: All students who are in the normally accepted terminal grade of the secondary school system and are studying mathematics as a substantial part (approximately 5 hours per week) of their academic program.

Such definitions could, of course, do no more than guide the choice of populations by National Centers. Terms that were necessarily ambiguous had to be defined in ways that were appropriate to individual school systems. Issues of feasibility had to be faced. In Finland and Israel, for example, linguistic minorities were excluded for reasons of cost from all or part of the Study. In Nigeria, only one region (the South) was included within the scope of the Study, again because of limited financial resources. The outcomes of the decision-making by the SIMS national committees about the ways in which the international guidelines for populations might be reflected within each system are set out in Tables 3.1.1 and 3.5.1. It can be seen that, although there are broad similarities among the various populations being investigated, there is some significant variation to be taken into account.

3.1 Population A

The population definitions for the 21 systems included in the curriculum analysis are set out below, together with a classification of the target grade using the conventions of the UNESCO *International Standard Classification of Education* (ISCED) (UNESCO 1981.)

Four aspects of these definitions will be considered here: the relationship of each system's target population to the grade in which students are found and the system's age cohort of 13-year-olds; the distribution of ages within

54

the samples chosen to represent these populations; the relation of the target populations to the grade structure of each system's school system; class size; and the number of courses within each system offered to the target population.

Table 3.1.1   *Population A: Definitions of Target Populations*

System	ISCED level	Population definition
Belgium (Flemish and French)	B2	Students in their second year of secondary schooling in both Type I (RSE) and Type II (Traditional) courses, from the general, technical and vocational streams.
Canada (British Columbia)	B1	All students in Grade 8, the first year of secondary schooling, in the public school system.
Canada (Ontario)	A8	Students enrolled in grade 8 mathematics courses, grade 8 being normally the last year of elementary education.
England and Wales	B3	Students in the third year in secondary schools of all types, or the equivalent year in middle schools where these exist, who reached their fourteenth birthdays in the academic year in which testing took place.
Finland	B1	Students receiving standard mathematics instruction at the grade 7 level (7. *Luokka*) in comprehensive or equivalent schools.
France	B3	Students in the third year of secondary education, (*Quatrieme*) in *collèges*, both public and private. Students who had transferred to a Professional Education *Lycée* (LEP) were not included.
Hong Kong	B1	All students in Form I, the first year of secondary schooling, taking mathematics as part of their curriculum.
Hungary	A8	All students in grade 8, the last year in the General School.
Ireland*	B1	All students in the first year of post-primary schooling.
Israel	B2	All students in Grade 8 in Hebrew-speaking schools, both State and State Religious, in both the Old and Reform systems.
Japan	B1	Students in grade 7 (*Chugakko 1 Nen*), the first year of lower secondary schooling, in public schools and national schools. Students in private schools and special schools (for handicapped students) were not included.
Luxembourg	B2	Students in normal classes in year 8 (*VI Secondaire, 8me Moyenne, 8me Professionnelle, 8me Complémentaire, 8me Secondaire Technique*), the second year of secondary education in each of the secondary school types in the system. Students in international schools were omitted.
The Netherlands	B2	Students in their second year of instruction at one of the following school types: VWO, HAVO, MAVO, LTO and LHNO. Students in other school types, comprising approximately 20 percent of the cohort, were excluded.
New Zealand	B1	Students in normal classes in Form 3, the first year of secondary education, in State secondary schools (including Coeducational, Boys', Girls', Forms 1–7 and Area Schools), and in Private and Integrated schools (including Coeducational, Boys', Girls').
Nigeria	B3	All students in Form III, the third year in secondary grammar schools, in the ten southern states of Nigeria.
Scotland	B2	All students in their second year (S2) at local authority or grant-aided secondary schools.

Table 3.1.1 (cont'd)   *Population A: Definitions of Target Populations*

System	ISCED level	Population definition
Swaziland	B2	All students in Form II in the system.
Sweden	B1	All students in grade 7 (*Årskurs 7*) of the 9-year compulsory school.
Thailand	B2	Students enrolled in grade 8 (*Maw 2*), the second year of secondary education under the revised national scheme of education of 1977, from both public and private schools.
United States	B2	All students in Grade 8 in mainstream public and private schools. For the majority of students, this is the second year of secondary schooling.

* Did not take part in testing.

3.2   Target Populations and Age and Grade Cohorts

The international guidelines for Population A envisaged that all students in the grade in which the modal age was 13 years would be included within the scope of the target population. It was further assumed that in most cases the *grade cohort* would include most if not all of the *age cohort*. (In all cases, however the scope of the term "all students" was restricted so as to exclude students who were in correctional institutions or hospitals or who had severe or moderate physical, intellectual or social handicaps.) However, in some systems, larger groups were excluded from the national population definition:

—Canada (British Columbia) (with an excluded population of 6 percent) did not include students outside the government-sponsored school system (i.e., students in "private" schools) or students requiring substantially-modified programs to suit their needs.
—France (with an excluded population of 12 percent) did not include students in "practical" or *aménagée* 8th grade classes and a small number of 13-year-old students who had been guided toward technical and professional training classes.
—Israel included only Hebrew-speaking schools within its target population.
—Japan excluded students in the private sector—under 5 percent of the age cohort.
—Luxembourg (with an excluded population of about 7 percent) did not test students in international schools.
—The Netherlands (reporting the exclusion of 20 percent of the age cohort) did not include a number of school types within their school system within the scope of the study. In some cases this was because the relevant students were at too low a level of attainment for the Population A test to be meaningful. In other cases it was because the curricula or schools concerned were of a mixed type and as such would not have added significant information for the analysis. The Dutch National Committee made the judgment that these excluded populations would not lead to serious bias in the estimates of the parameters.

In each of these cases the excluded population is described in terms of the *age cohort*. In other cases, however, the pattern of exclusion is more complicated and estimates of the scope of the excluded population have proven more difficult to obtain. Thus, Nigeria excluded from its target population pupils enrolled in trade schools, technical and other vocational and prevocational schools, and students in schools that had been established for less than 5 years. Furthermore, as has already been mentioned, limited resources restricted the study to 8 of the 10 southern states of the nation.

Moreover, Nigeria, like Swaziland and Thailand, does not enroll all 13-year-olds within the secondary school system. The within-school population is estimated as representing 33 percent of Thailand's 13-year-old age cohort. Estimates of grade and age cohort relationships are not available for Nigeria and Swaziland.

In summary, then, while most participating systems came very near to including "all students" in school within their target population (as this goal is mediated through a grade cohort) a number of national centers omitted parts of cohorts from their target populations. In some cases it may be that the effects of these patterns of exclusion have resulted in a biasing upwards of both curricular expectations and achievement.

3.3 Placement of Population A Within School Systems

It is a truism that the internal organization of school systems varies and that curriculum content is deployed differently across the grades of different systems. In order to minimize the consequences of such potential heterogeneity, the guidelines for SIMS suggested that the target grade for Population A be defined in terms of a modal age. But it remains an open question whether or not the grades actually selected by the National Committees represent comparable points across all the participating systems. Two questions emerge as consideration is given as to whether or not the guidelines were able to unify the different populations around a common curriculum stage:

1. Are the different Population A's comparable in terms of the *ages* of students?
2. Are the different Population A's comparable in their *place* within their systems?

Table 3.3.1 reports the ages of students in the *achieved* Population A samples for all systems included in the curriculum analysis except Ireland (which did not participate in the testing of the study). The data are drawn from the background questionnaires that were completed by all students involved in the individual Population A samples.

It is clear that two cases, Nigeria and Swaziland, are markedly different from the other systems in the ages of the students in their samples. In each case the modal age was 15 or 16 years at the time of testing and many students were considerably older. Among the other systems, Luxembourg, French Belgium and The Netherlands appear to have tested students older than 13 (median age 172 months or 14.4 years at the time of testing). The students in Japan and Hong Kong were younger than those in the other systems, with medians of 162 and 157 months respectively—the result of decisions by those systems about the appropriateness of the SIMS item pool to national curricula.

Table 3.3.1   *Population A: Distribution of Student Ages in Achieved Samples*
(in months)

System	Median	Mean	Standard deviation
Belgium (Flemish)	169	171	8.0
Belgium (French)	172	174	11.3
Canada (British Columbia)	167	168	6.0
Canada (Ontario)	168	169	6.8
England and Wales	169	170	3.8
Finland	166	166	4.8
France	169	170	8.3
Hong Kong	157	159	10.9
Hungary	170	171	13.4
Israel	168	168	4.7
Japan	162	162	3.5
Luxembourg	173	175	8.9
Netherlands	172	173	7.9
New Zealand	168	168	5.4
Nigeria	192	200	37.7
Scotland	168	168	4.3
Swaziland	185	188	22.5
Sweden	167	167	4.2
Thailand	171	171	9.0
United States	169	170	6.0

To consider the comparability of the *grades* included in the different Population A's, it is useful to refer to the UNESCO (1981) *International Standard Classification of Education* (ISCED) which attempts to standardize at least the major features of school organization for the purposes of international comparisons. One such feature is the point of transition between elementary, primary, or First Level schooling and secondary or Second Level schooling. It is here that many features of the school experience of students change (classroom organization, selection into school types and ability groups, etc.) and it would seem that the First Level-Second Level divide might be a useful starting point for determining where Population A for each system is located in terms of a common point of reference.

Table 3.3.2 indicates the number of years before and after the First Level-Second Level divide in the various Population A grades. The meaning of this divide must be somewhat ambiguous in that this organizational break point is not necessarily closely calibrated with curricular organization. In the United States, for example, grade 8 is the second secondary year in most jurisdictions, but in terms of the organization of mathematics it is often the last year of general "elementary" mathematics, the one preceding the first course in algebra. However, despite such issues, it is clear that the various Population A's span a range of grade levels around the First Level-Second Level Divide, from 1 year *before* in the case of Hungary, Israel (nonreformed), Ontario, and the U.S. to 3 years *after* in the cases of France, England and Wales, and Nigeria. There is a strong indication in these findings that, despite the similarity in underlying age groups, *systems are being sampled at somewhat different points in their internal structures.* This possibility has many implications for the Study and, in particular, it offers some clues about the circumstances that might lie behind the patterns of content coverage found within different participating systems. This issue will be pursued further in Chapters 4 and 5.

Table 3.3.2 *Place of Population A in the School System*

System	Grade after the first level-second level divide
Belgium (Flemish)	2
Belgium (French	2
Canada (British Columbia)	1
Canada (Ontario)	−1
England and Wales	3
Finland	1
France	3
Hong Kong	1
Hungary	−1
Ireland	1
Israel	−1 (in nonreformed) 2 (in reformed)
Japan	1
Luxembourg	2
Netherlands	2
New Zealand	1
Nigeria	3
Scotland	2
Swaziland	2
Sweden	1
Thailand	2
United States	−1

Notes:
Positive numbers refer to Second Level grades.
Negative numbers refer to grades before the Second Level.

3.4 Class Size (Population A)

Class size is widely thought of as a major influence on educational achievement and could, in the context of SIMS, be as important a determinant of achievement and teaching practices as any of the factors being considered in this chapter. How similar are the systems in the sizes of their classes at the Population A level?

Table 3.4.1 presents the median and mean class sizes in the target populations for each system together with the standard deviations; the data was derived from the *Teacher Questionnaires* completed for each participating class. As can be seen in the Table, there is substantial between-system variation in the size of the modal class; from 19 students in the case of Luxembourg to 41 in the cases of Japan and Swaziland and 43 in the case of Thailand. Most systems, however, have class sizes in the 20s and it is the classrooms in the low- and middle-income systems of Hong Kong, Swaziland and Thailand which tend to be the largest. Israel and Nigeria have the greatest within-system range in class size.

Table 3.4.1   *Number of Students in Target Class (Population A)*

Country	Median	Mean	Standard deviation
Belgium (Flemish)	21	20.3	5.4
Belgium (French)	20	19.7	5.2
Canada (British Columbia)	28	26.9	4.6
Canada (Ontario)	29	28.5	5.1
England and Wales	27	25.4	8.2
Finland	22	22.0	7.0
France	24	23.8	3.0
Hong Kong	44	43.3	4.3
Hungary	26	25.8	5.3
Israel	22	23.7	10.4
Japan	41	39.4	6.4
Luxembourg	19	19.0	5.8
Netherlands	24	23.9	3.5
New Zealand	29	28.3	4.4
Nigeria	30	32.8	14.8
Scotland	29	27.6	5.7
Swaziland	41	37.8	6.5
Sweden	20	19.9	5.7
Thailand	43	42.0	7.1
United States	26	26.4	7.0

3.5   The Curricular Organization of Population A

Section 3.3 concluded that the various Population A's may represent somewhat different points within the school organization of the various participating systems. In this section, variation of the underlying organization of mathematics instruction within the different systems will be considered. This section will focus on the formal organization of instruction, i.e., the allocation of time to instruction in mathematics and the number and character of the "courses" that make up Population A mathematics.

The most basic organizational variable within schools that affects curricula, instruction and achievement is the time allocation given to subjects. Table 3.5.1 presents the means and standard deviations of hours per year devoted to mathematics instruction as reported on the *Teacher Questionnaires*. Table 3.5.1 also gives the percentage of the total timetable devoted to mathematics as reported by National Research Coordinators.

A cursory inspection of Table 3.5.1 suggests that while many systems allocate 120–140 hours to mathematics at the Population A year, there is substantial variation around this modal allocation. Several systems devote less than 100 hours annually—Finland (83 hours), Hungary (77 hours), Japan (99 hours), Sweden (96 hours) and Thailand (96 hours). Other systems devote more than 145 hours to the subject—Luxembourg (147

Table 3.5.1   *Population A: Number of Hours of Mathematics Instruction and Percentage of Total Timetable*

Country	Mean hours	SD	Percent of timetable*
Belgium (Flemish)	140	33.0	9
Belgium (French)	140	29.1	9
Canada (British Columbia)	128	24.6	14
Canada (Ontario)	140	33.6	12
England and Wales	117	24.8	13
Finland	83	4.5	10
France	129	12.0	13
Hong Kong	124	39.8	16
Hungary	97	6.6	13
Ireland	110†	–	12
Israel	133	26.4	10
Japan	99	15.2	12
Luxembourg	147	26.9	10–16
Netherlands	121	40.9	7–13
New Zealand	132	24.3	13
Nigeria	144†	–	10
Scotland	148	21.7	17
Swaziland	157	58.8	–
Sweden	96	0	11
Thailand	96	38.9	13
United States	146	35.0	15

* Percentage provided by National Research Coordinators on the basis of the intended curriculum. Percentages may vary by program and/or course.
† Estimate provided by National Coordinator.

hours), Scotland (148 hours), Swaziland (157 hours) and the United States (146 hours). These differences reflect factors such as the length of the school year. They also reflect national differences in the importance given to mathematics relative to other subjects. Thus, in systems such as Flemish and French Belgium, Finland, Ireland, and Sweden, where second national languages or foreign languages are given substantial attention at this level, there is less time available for other subjects. It can be predicted that such differences in both relative and absolute time would have discernible consequences for content coverage (and student achievement). For the intended curriculum, both the scope of coverage and the depth of coverage of given topics would be expected to be greater in systems where there is more available time for instruction. For the attained curriculum it might be suggested that, all other factors being equal, student achievement would be higher in systems that devote more time to mathematics instruction.

While time allocation is one crucial dimension of curricular organization, course structure is another in that it determines the ways in which content is deployed across classes. Again, understanding of these patterns and structures is important when determining how systems' responses to the SIMS table of specifications and international item pool are to be interpreted. In some cases, it might be inferred, such responses would refer unambiguously to all students in a system; in other cases such responses must be seen as referring to particular courses and the groups of students who take those courses. A given item may be important for some students within a system but not for all.

Table 3.5.2 sets out the number of courses offered at the Population A level that represent *coverage of effectively-different content.* An examination of the table shows that one-half of the nations participating in SIMS offer only one course at the Population A level but that some systems offer 3 or 4 different courses. The pattern of distribution of students across such courses also varies. In some cases students are more or less equally divided among the available courses (England and Wales, Hungary, Luxembourg, The Netherlands) while in other cases students tend to congregate in one or another of the available options (Finland, Scotland, and Sweden).

Table 3.5.2   *Population A: Number of Effectively-Different Courses and Percentage Taking the Most Popular Course*

System	Number of effectively different courses*	Percentage taking most popular
Belgium (Flemish)	2	81
Belgium (French)	2	85
Canada (British Columbia)	1	100
Canada (Ontario)	1	100
England and Wales	3	33
Finland	2	78
France	1	100
Hong Kong	1	100
Hungary	3	46
Ireland	1	100
Israel	1	100
Japan	1	100
Luxembourg	5†	35
Netherlands	4	33
New Zealand	1	100
Nigeria	1	100
Scotland	2	90
Swaziland	1	100
Sweden	2	74
Thailand	1	100
United States	4	66

* Note: "Effectively different" refers to substantial differences in content and/or goals. Courses with different titles but very similar in content are not distinguished in these data.
†In Luxembourg, five different programs were available at the time of data collection. This number has since been reduced to four.

Again, these differences in both the number of courses available within a system and the composition of the course-taking patterns of subpopulations would seem to have consequences for both coverage patterns and achievement profiles. But while this reflection is broadly true, the actual character of the course differentiation found within and between nations will also affect the character of any profiles that might emerge. And while systems may have emphasized one slice of their curricular reality in describing their course structures, other realities are also present. England and Wales, for example, distinguished courses on the basis of mathematical content ("traditional," "modern," "compromise") but within these types (which are themselves abstractions from the many courses actually taught) ability banding or grouping is commonplace. Finland, Luxembourg, The Netherlands and Sweden, on the other hand, differentiate courses to reflect aptitude and achievement differences among students but within this structure there are differences in content to be found. This is also true for the United States where, within one diffuse national structure different courses organized on the basis of both levels of content and coverage are targeted on subgroups of varying aptitude and attainment.

3.6 Population B

The population definitions for the 18 systems taking part in the curriculum analysis at the Population B level are set out below. Four aspects of these definitions will be considered: there will be a brief discussion of Population B in general and of some of the problems experienced by individual systems in defining their national populations. Following this, there will be an exploration of the relationship between the populations and age and grade cohorts. There is then an examination of the time allocations represented by and class sizes in the "courses" that define the Population B's. Finally, the number of courses included within the scope of each system's definition of Population B is considered.

Table 3.6.1 *Population Definitions—Population B*

System	Population definitions
Belgium (Flemish and French)	Students in the last year of secondary school, receiving at least 5 hours of mathematics instruction per week.
Canada (British Columbia)	All grade 12 students enrolled in the course Algebra 12 in the public school system.
Canada (Ontario)	Students enrolled in two or more grade 13 mathematics courses (Calculus, Relations and Functions, and Algebra).
England and Wales	Students in the second year sixth form studying at least one mathematics course for the General Certificate of Education (GCE) at Advanced-level or Scholarship-level. Students in sixth form colleges or in independent schools were included in the population.
Finland	Students taking the long course in mathematics in grade 3 of Finnish-speaking upper secondary schools. Swedish-speaking upper secondary schools, and upper secondary evening classes were excluded.
France*	Students in *Terminale* C (mathematics and physics) and *Terminale* D (mathematics and natural sciences), two tracks of the last year at a public or private *lycée*. Students in both classical and polyvalent *lycées* in metropolitan areas were included.
Hong Kong	Two populations were defined at this level: two preuniversity years exist, with different characteristics in terms of curriculum, age of students and language of instruction (Chinese or English).

Table 3.6.1 (cont'd)   *Population Definitions—Population B*

System	Population definitions
	Population B1 consists of students in Form Lower Six, the course for those planning to enter the Chinese University of Hong Kong. *Population B2* consists of students in Form Upper Six, those destined for entry to the University of Hong Kong. These students have typically had one more year of secondary education than those in Population B1.
Hungary	All students in the fourth grade of grammar, specialized vocational, and technical schools.
Ireland*	All students in their Leaving Certificate year taking mathematics at the Ordinary or Higher level.
Israel	Students in the terminal grade in those Hebrew-speaking schools offering at least 4-point mathematics programs, a "point" representing 90 learning "periods" during grades 9–12.
Japan	Students in grade 12 *(Kotogakko 3 Nen)* of the fulltime upper secondary schools, public, national and private, who are studying mathematics as a substantial part (more than 5 hours per week) of the general course or science-mathematics course.
Luxembourg*	Students in the year 13 *(I Secondaire)* in the mathematics, natural science and economic science sections of the *lycée*.
New Zealand	Form 7 students in state secondary, private and integrated schools (including Coeducational, Boys', Girls'), who are studying pure mathematics as a substantial part of their programs (approximately 5 hours a week).
Scotland	All students in their fifth or sixth years at local authority or grant-aided secondary schools preparing for one of the following examinations: SCE Higher Mathematics, GCE Advanced Level Mathematics, Scottish Certificate of Sixth Year Studies in Mathematics.
Sweden	All students taking the common grade 3 mathematics course in the Natural Science line and the Technical line of the upper secondary school *(Gymnasieskölan)*.
Thailand	Students following the mathematics/science program in their last year *(Maw 6)* of upper secondary education at a public or private academic school.
United States	Students in mainstream public and private secondary schools who are enrolled in a mathematics course having as its prerequisite the standard sequence of mathematics courses: Algebra I, Geometry, and Algebra II.

* Did not take part in testing at this level.

The placement of Population B within their respective school systems does not need to be considered in detail because, by intention, Population B was the population of "mathematics specialists" in the *terminal grade* of each participating school system. Nevertheless, some comparability of expectations across these terminal grades is assumed and it is important to test the limits of this comparability at both the structural and the curricular level. This latter question will be explored in Chapters 4 and 5. In the present chapter, some structural data will be considered as a basis for a preliminary assessment of the extent to which the guidelines for the study did unify the set of Population B's.

In most cases the choice of a target population posed few problems for national committees. However, there were two systems in which the choice of the target population posed major problems for national committees. The Population B "grade" that was finally selected spanned two grades within their school systems. In Hong Kong there are 6 years in the Chinese secondary school system and 7 years in the English system. Both groups of students were included in Population B. In Scotland, the 5th and the 6th years of secondary school can both be considered terminal and, as a consequence, the national Population B was drawn from both groups.

Table 3.6.2 reports the age of students in the achieved samples for Population B for the 15 systems included in the Population B testing. For most systems the median ages cluster in the range 214 to 220 months (17.1 to 18.4 years). In Scotland, however, the median age is 201 months (16.9 years), fully 13 months less than the median age for any other system. In Finland, Ontario and Sweden the median ages tend to be somewhat higher. In general, however, it would seem that the Population B's do encompass students who are of similar age.

Table 3.6.2   *Population B: Distribution of Student Ages (in months)*

System	Median	Mean	Standard deviation
Belgium (Flemish)	216	217	10.2
Belgium (French)	217	220	11.1
Canada (British Columbia)	215	215	6.3
Canada (Ontario)	225	222	13.9
England and Wales	217	218	3.9
Finland	223	223	6.3
Hong Kong	220	222	12.3
Hungary	217	217	4.2
Israel	214	214	4.8
Japan	217	218	4.0
New Zealand	214	214	6.0
Scotland	201	202	6.8
Sweden	228	230	18.7
Thailand	218	218	9.0
United States	214	213	7.2

A more substantial problem affecting the comparability of the Population B's derived from the definition of the target population *within* the school system. In contrast to Population A, where the intention was the inclusion of "all" students within the designated grade, the international guideline for Population B stated that the population should consist of those students who take mathematics as "a substantial part (approximately 5 hours per week) of their academic program." In some, if not most, educational systems, this guideline was easily interpreted. In others, various difficulties were encountered as the guideline was applied.

There were at least three types of situations to be considered. In the first, the guideline is easy to interpret: the educational systems have clearly identifiable mathematics tracks or offer substantial mathematics courses taken by those who selected them as a major element of their studies. This is the case in France and Sweden where the final year's instruction is organized into distinct "lines" or tracks and only students enrolled in mathematical, scientific and (in Sweden) technical lines were chosen. A similar situation exists in England and Wales, Hong Kong, and New Zealand where students take only a few major subjects, each forming a substantial part of the program, with the implication that all students studying one or more terminal year mathematics courses are specialists. The second case involves systems where mathematics is one of many optional subjects that students are genuinely free to take or drop. Classes in mathematics in such systems enroll some students who hope to specialize in mathematics or mathematics-based programs at the university level and others who take the subject out of interest or availability although their main concerns are elsewhere. The North American curriculum tradition—as seen within SIMS in Canada (British Columbia) and the United States—tends to operate in this manner. Thirdly, there are some systems in which a substantial mathematics course either is compulsory at the Population B level (as in Hungary) or is required for so many occupations and programs in higher education that very few students drop it (as in Ireland).

Needless to say, there are systems whose curricular patterns are not easily classified in these terms but these distinctions provide an important way of looking at the characteristics of the Population B programs and students. In the case of the second and third of these categories, with their wider definitions of mathematics course-taking, the population of specialist mathematicians is almost by definition larger than it is in the first case. This broadening of the Population B definition has obvious implications for curriculum and achievement.

3.7   Population B in the Context of Age and Grade Cohorts

One obvious consequence of the curricular traditions that lie behind the kind of categorizations of school mathematics sketched above is a set of differences between systems in the proportion of the *grade* cohort that makes up the various Population B's. And, of course, these grades are embedded within school systems which enroll varying proportions of the *age* cohorts.* The result, as seen in Table 3.7.1, is a set of immediate contexts for

---

* Some school systems, e.g., Canada (British Columbia), Japan and the United States, define "school" expansively at the upper secondary level and embrace many forms of vocational training with the secondary school. Other systems are much more restrictive in their definition of "school" and tend to classify many training programs as "further education" and the like. In Table 3.7.1, Column 2 accepts each system's own definition of the scope of the "school" at this level, the definition was used in determining the SIMS sampling frame; column 3 embraces all forms of formal education and training of Population B age pupils within a definition of "education and training."

the various Population B's that vary markedly. As a proportion of the *age* cohort Population B varies from 50 percent in the case of Hungary to 6 percent in the cases of England and Wales and Israel.

Table 3.7.1 *Population B: Size of Target Population as a Percentage of Age and Grade Cohort*

System	Pop B / age cohort	Students in school/ age cohort	Students in education and training/ age cohort
Belgium (Flemish)	10	60	60
Belgium (French)	10	60	60
Canada (British Columbia)	30	82	82
Canada (Ontario)	19	33	49
England and Wales	6	19	56
Finland	15	59	59
Hong Kong B1	8	17	17
B2	4	7	7
Hungary	50	50	50
Ireland	50	55	55
Israel	6	60	60
Japan	12	92	92
New Zealand	11	17	43
Scotland	18	36	57
Sweden	12	24	42
Thailand	NA	17	NA
United States	13	82	82

NA = Data not available.

In summary then it is clear that there are very different contexts for Population B. These range from cases in which Population B consists of a small group of students within a comparatively restricted (i.e., in the terms of the proportion of the cohort being enrolled) secondary school system— England and Wales—to cases in which Population B represents a large group of students in a school system in which most students are enrolled— Canada (British Columbia). It follows that there are differences to be expected in the scope of the intended, implemented and attained curricula for Population B in these diverse contexts.

3.8 The Curricular Organization of Population B

Differences in the context of Population B also emerge in the organization of curricula in the various systems and in the allocation of time for instruction. Some of the differences between the various populations are even more striking at this level than they are in Population A. Table 3.8.1 presents the time allocations of mathematics instruction for the sample courses. (The time estimates were reported by teachers in the achieved sample and the estimates of the proportion of the timetable devoted to single courses in mathematics were reported by National Research Coordinators.)

In Finland, to take the extreme lower bound, Population B mathematics is given 66 hours in the terminal year.* In Flemish and French Belgium, England and Wales, Hong Kong, Israel, Scotland, Thailand and the United States the course or courses defining the Population B target groups are given more than 150 instructional hours. In Canada (British Columbia), Finland, and Hungary, an advanced mathematics course represents less than 15 percent of the timetable for those students taking it. In England and Wales, Hong Kong and Ontario such courses take up over 22 percent of the timetable.

Moreover, in those nations whose curricula have been influenced by the English educational tradition, it is a common practice for many students who see themselves as specialists in mathematics and science to take two or more concurrent courses in mathematics in the terminal year. In such nations (Canada (Ontario), England and Wales, Hong Kong, New Zealand), the time allocations and the proportions of the timetable reported in Table 3.8.1 can be at least doubled for a significant number of students. Needless to say, if time is the principal resource available for instruction, very different situations are present in Population B mathematics.

Table 3.8.1   *Population B: Hours of Mathematics Instruction Per Year and Percentage of Total Timetable of One Mathematics Course*

	Mean hours	SD	Percent of timetable*
Belgium (Flemish)	202	45	18
Belgium (French)	195	52	18
Canada (British Columbia)	122	25	14
Canada (Ontario)	112	16	25
England and Wales	180	66	22
Finland	67	13	12
Hong Kong	182	91	25
Hungary	109	40	15
Ireland*	130	–	13
Israel	150	35	20
Japan	149	37	15
New Zealand	139	19	15
Scotland	174	23	20
Sweden	120	0	16
Thailand	144	54	21
United States	150	24	16

Source: Teacher questionnaire
* Estimate provided by National Research Coordinator

This envelope of curricular organization also affects the kinds of internal differentiation of mathematics courses and instruction within systems. Table 3.8.2 reports the number of "effectively-different" courses offered to students *within* the Population B target groups of the participating systems,

---

* In Finland, teaching in the Population B grade ends early in the school after which point students prepare for final examinations. In the previous grade, mathematics instruction is allocated 128 hours.

together with the estimates provided by National Research Coordinators of the proportion of the target population taking the most commonly taught of these courses.

Table 3.8.2 *Population B: Number of Effectively-Different Courses Offered Within the Target Population*

	Number of different courses	Percent taking most widely-taught course	Percent taking more advanced topics
Belgium (Flemish)	4	41	57
Belgium (French)	2	60	40
Canada (British Columbia)	3	100	16
Canada (Ontario)	3	95	36
England and Wales*	4	N.A.	–
Finland	1	100	50
Hong Kong	4	81	–
Hungary	3	93	7
Ireland	2	90	10
Israel	1	100	–
Japan	1	100	–
New Zealand	2	100	40
Nigeria	1	100	–
Scotland	4	79	–
Sweden	1	100	–
Thailand	1	100	–
United States*	5	70	30

* No single course prescription exists. The number given relates to course types.

Again, even a cursory inspection suggests the range of course-providing patterns found in the participating systems. In Canada (British Columbia), Japan, Sweden and Thailand all Population B students take what is either actually or in effect one course; in Flemish and French Belgium, Canada (Ontario), England and Wales, Hungary, Scotland and the United States the Population B target groups include students who are two or more effectively-different courses. Furthermore, as can be seen in Table 3.8.2, the number taking the most widely-taught of these courses ranges from 40 percent in the case of Hungary to 70 or more percent in the case of the United States and Scotland.

Table 3.8.2 reports the number of courses which can be identified as constituting separate courses seen in terms of the *formal* provision of courses to Population B students by the school system. In addition, there is, in some systems at least, a variety of possibilities for students, teachers and schools that can be created out of the options made available by their systems. In Finland, for example, supplementary "short" courses covering a wide variety of topics are made available for terminal year students and 50 percent of students take one or more such extra "courses." In New Zealand, approximately one third of Population B students take a course in applied mathematics *in addition to* the course in pure mathematics that was used to define the Population B target group. In Canada (Ontario), England and Wales and Hong Kong similarly heterogeneous patterns are to be found.

The plethora of course combinations that can be created has, significant implications for descriptions of national curricula. As decisions of individual students aggregate, patterns are produced based on intersecting sets defined by individual courses and their possible combinations. At times there may be considerable overlap in the content and skills emphasized. At other times the content of individual patterns may become a discrete "course" quite different in character from other "courses." But as all such combinations become patterns, they create realities that can seriously qualify any blanket statement about what a given system's curriculum might be. In New Zealand, for example, probability and statistics is a topic within the Applied Mathematics course. Students who take that course in combination with Pure Mathematics—the defining course for membership of New Zealand's Population B—are likely to learn far more about these topics than are their peers taking Pure Mathematics alone. Such possibilities need to be considered when interpretations are being made of coverage indices and patterns of achievement within a system at the Population B level.

It is possible to sketch five basic patterns of course and curriculum organization at this level, each of which would seem to have different implications for national patterns of coverage and achievement variance.

1a. *One course only is offered to Population B students*. This is the case in Canada (British Columbia), Israel, Japan, Nigeria, Sweden and Thailand.

1b. *One basic course is offered Population B students, but it may be augmented* by one or more formally-designated "courses" which may cover different or more advanced material. Such "courses" may be "short courses" (as in the case of Finland) or "regular courses" taken by some students (as in the case of Canada (Ontario) and New Zealand).

2. *A number of discrete courses can be offered* Population B students representing (a) different levels or (b) kinds of content, but students take only *one* of these options.

Situation (a) is found in Hungary while (b) occurs in French and Flemish Belgium, Scotland and the United States.

3. *A number of courses and course types and various combinations of these are available; students may select one or more different patterns*. This is the case in England and Wales and Hong Kong.

Needless to say, it must be assumed for the purposes of analysis and evaluation that different national philosophies and/or ideologies of mathematics education lie behind these different forms of curriculum organization. In the case of (1b), (2) and (3) *intended diversity* is seen, and appropriate variations in within-system implemented coverage and, eventually, student attainment are expected. Such expectations are not held for nations organizing their curriculum in the terms suggested by (1a).*

3.9 Class Size (Population B)

The previous discussion has suggested that there is marked variation among the systems which participated in the Population B component of SIMS in the context of advanced mathematics classes, curricular organization and, by implication at least, overall aims and intentions for advanced mathematics at the upper secondary level. To what extent do these differences manifest themselves in class size? As we have already noted, class size is widely regarded as a major influence on educational achievement and it clearly plays a significant role in determining the environment of teaching.

Table 3.9.1 presents the findings on the mean size of classes in the Population B target population for the systems which participated in the testing phase of the study and, again, we see substantial between-system variation—and more variation than is seen at the Population A level. Median class size ranges from 10 in the case of England and Wales to 40 in the case of Thailand and 43 in the case of Japan. Two systems have median class sizes below 15 (Belgium (French) and England and Wales), seven systems have class sizes between 15 and 25 and four systems have class sizes over 25. Thailand has the greatest variability in class size (SD 15.0).

Table 3.9.1 *Number of Students in Target Class (Population B)*

Country	Median	Mean	Standard Deviation
Belgium (Flemish)	16	15.5	8.8
Belgium (French)	13	13.9	7.3
Canada (British Columbia)	25	23.4	7.7
Canada (Ontario)	25	24.6	7.2
England and Wales	10	10.2	5.3
Finland	20	19.5	6.4
Hong Kong	27	27.3	10.1
Hungary	27	26.4	7.3
Israel	16	18.1	9.9
Japan	43	40.4	8.7
New Zealand	16	15.8	5.4
Scotland	23	21.9	6.7
Sweden	23	22.4	5.1
Thailand	40	43.0	15.0
United States	20	21.6	9.7

3.10 Curriculum Control

We have been suggesting that there are some very basic differences in the ways in which the school systems participating in SIMS are organized—with

---

* In addition, in the cases of Hong Kong and Scotland the Population B target populations include groups of students drawn from different grade levels within their school systems. Different mathematics courses are offered at these levels.

perhaps the extreme contrast being between the systems in which there is no single school system except in the abstract sense (e.g., the United States) and those nations in which there is a tightly-organized national system (e.g., Sweden). In this section we will consider two particular aspects of this cluster of possible differences between systems, the forms by which curricula and teaching are developed, supervised and controlled and that by which "standards" of coverage and attainment are monitored.

We have already discussed the extent to which curriculum planners in a school system *intend* there to be differentiation and diversity of content coverage with their school systems—an inevitable outcome of the provision or nonprovision, or availability or nonavailability, of a variety of courses within a school system. However, our development of this theme in Sections 3.4 and 3.7 did not emphasize the *forms* of availability of such options. In some cases, such course differentiation comes about as a result of interaction between, on the one hand, a set of course types (operationalized as both models and textbooks) and, on the other hand, choice by local authorities, schools and teachers about which of this potential set of course-types they will offer. In this kind of case (and we see this situation most clearly in England and Wales and the United States) a statement of the number of courses "available" at a given level is, at bottom, an empirical statement: a judgment about how many *course-types* or *course-families* are found in the schools. In other cases it is *courses* in the narrow sense which are prescribed or mandated for different subpopulations within the school. Thus in Finland and Sweden the central educational authority mandates the availability of two courses for students at the Population A level and while the decision about who might enter one of the other such courses is left to individual students and their families, the coverage and the standards of expectation and attainment within the courses are seen as fixed by a central authority.

Within these two broad classes of course development and availability and prescription (or nonprescription) there is also a set of broad differences between school systems in the extent to which they give local education authorities, schools and teachers the opportunity to interpret how they will present courses, what they will emphasize, what standards of accomplishment they will seek to maintain, and what texts they will use. Thus, in Canada (Ontario) there have been shifting emphases over the past few years in the extent to which the central educational authorities intend central mandates to be authoritative over, or merely suggestive to, local authorities or schools—and such oscillations have affected the educational culture of the system in important ways. In England and Wales, on the other hand, schools or local educational authorities have been traditionally free to develop their own courses in their own ways or, more accurately, to choose which of several available course families they will adopt—but all such families are firmly articulated with an examination system that will be the basis for determining ultimate "success" and, therefore, the implied standards of accomplishment. In the United States there is no such articulation of course families with a nationally-effective examining system. Furthermore, there are few constraints on the ways in which teachers and schools might interpret diffuse notions about what is appropriate for a given grade or class level.

Table 3.10.1 offers a summary characterization of the "level" (defined as overall jurisdiction, i.e., "nation," region or state, local education authority, and school) at which decision-making on content and standards is formally vested in the systems participating in SIMS. The unbracketed letter in each column represents a judgment reported by the National Research Coordinator about the vesting of these decisions for "academic" curricula; a letter in parentheses represents the situation for nonacademic curricula where there is explicit provision for the less able students. As is indicated by an inspection of the entries in the first column, in most systems decision-making about content is a firm system responsibility. Only in the cases of England and Wales and the United States do we see decision-making being vested (theoretically) entirely in local authorities and schools. In New Zealand and Scotland, however, local authorities and schools do have an important role; and in the Netherlands there is explicit school-based decision-making about courses that are offered lower capability students.

Table 3.10.1 *Population A and Population B: Levels of Curricular Decision-making*

System	Content	Standards
Belgium (Flemish)	C	C,S
Belgium (French)	C	C,S
Canada (British Columbia)	C	S,C
Canada (Ontario)	C	S
England and Wales	S	C(S)
Finland	C	S,C
France	C	C
Hong Kong	C	C,S
Hungary	C,L	C,L
Ireland	C	C
Israel	C	C,S
Japan	C	C
Luxembourg	C	C(S)
Netherlands	C,S	C,S
New Zealand	C,S	C(R)
Nigeria	C(R)	R
Scotland	C,L,S	C(S)
Swaziland	C	C
Sweden	C	C
Thailand	C	S,C
United States	L,S	S

Key:
C—System
R—Region
L—Local Authority
S—School
Parentheses () indicate the practice for lower ability students.

The first column of Table 3.10.1 deals with *intended* curriculum; the second column deals with how the curriculum is implemented, and how student attainment is evaluated. Thus, while the first column reports the formal (or "legal") location of control, the second is concerned with how it functions in practice. And as we turn to the second column, we see that the curricular freedom found in England and Wales, and the relative freedom in Scotland and in New Zealand, are in fact limited in practice by national examination systems, in the upper secondary school at least. (Courses for less academic students may not be so affected.) But many other countries make provision, within the structures of a standardized national curriculum, for discretion in the interpretation, of both the scope of the curriculum and the required standards of accomplishment (both interim and final school certification, school graduation, university admissibility, and so forth). The ways in which such discretion is exercised can vary substantially from system to system. However, in Flemish and French Belgium, Canada (British Columbia), Canada (Ontario), Israel, and Thailand, schools play more or less significant roles in determining the final certification of Population B students; this is indicated by the presence of an unbracketted *S*, either with or without a *C*, in the second column of Table 3.11.1. Elements of local control are, of course, more widespread at the Population A level. How extensively such opportunities are exploited by schools and teachers depends on local tradition and culture; it is clear, however, that in at least some countries the degree of freedom assumed by schools—in interpreting the scope and implications of a national curriculum and the appropriate standards of accomplishment—is substantial.

As has already been indicated, in the case of students enrolled in curricula that are *not* oriented toward possible university entrance, discretion is more likely to be vested in the school. This is particularly true as regards standards, but applies also in some countries to the choice of content for such students. Part of the reason is, of course, the absence of an orientation toward the expectations of upper secondary school and university, and presumed lack of relevance of the backwash from certification and coverage concerns of the upper secondary school. And it is, we suspect, the practices of upper secondary school, and particularly the terminal year, which has the most significant consequences for limiting the presumed discretion of teachers, even at the lower secondary level. Thus, in the case of England and Wales, decision-making about curricula is formally vested in most respects in the school, but the *consequences* of such discretion are profoundly modulated by the requirements of the elaborate system of formal and external examinations found in England and Wales. Conversely, in Canada (British Columbia) and Canada (Ontario), we note that the curricula to be taught in the schools are firmly prescribed by the provincial (central) authority; but, in each case, the absence of firm systems to control, monitor and supervise the implementation of these curricula and attainment within them likewise modulates the seeming consequences of such central prescription.

Table 3.10.2 presents yet another interpretation of the pervasiveness of central or national curricular control; it focuses on the extent to which mechanisms for centralized supervision and central assessment of the accomplishments of students in academic curricula are to be found in the systems participating in SIMS. As can be seen, the classification recorded in the table confirms the general pattern identified to this point. In most systems, the work of students and teachers in the upper secondary school is directed towards standards prescribed by one or another form of external authority. But Canada (British Columbia), Canada (Ontario) and the United States rely almost totally on the judgments of teachers and schools for the certification of the accomplishment of their students.* We place in the middle of our spectrum those systems in which the school can make a contribution to determining the eligibility for certification or graduation.

Table 3.10.2 *Population A to Population B: Control of Coverage and Standards*

---

1. External control of curricular coverage and standards for Population B students. Mechanisms may include "external examinations," school entrance examinations when these are firmly based on national syllabi, university admission examinations and tests, etc. (The number of such examination points in the career of university-bound students is shown in parentheses.)

England and Wales	(2)
Finland	(1)
France	(1)
Hong Kong	(2)
Hungary	(2)
Ireland	(2)
Japan	(1)
Luxembourg	(2)
Netherlands	(1)
New Zealand	(2)
Nigeria	(2)
Scotland	(2)
Thailand	(2)

2. Formal external monitoring of school-based assessments. (The number of such monitored points is shown in parentheses.)

Belgium (Flemish)	(2)
Belgium (French)	(2)
Israel	(1)
Sweden	(3)

3. Customary control of standards: grading standards are largely determined at the local or school level.

Canada (British Columbia)
Canada (Ontario)
United States

---

* It is important to note as we interpret data on school credentialling structures that increasing numbers of systems have different requirements for secondary school graduation than those associated with university admission. Thus in Finland, Israel (for some programs), Thailand and some sectors of Flemish and French Belgium, and the United States systems additional examinations or tests are required for university admission of some or all prospective students. Such tests have their own backwash effects on the school. Canada (British Columbia) has a set of "scholarship" examinations at the Grade 12 level which also has important effects on school practices.

3.11   Curricular Diversity: A Typology

It should be obvious from the discussion in the preceding sections of this chapter that the systems participating in SIMS differ markedly in their patterns of curricular offerings and in the scope and pervasiveness of the control they impose on their schools. We can presume that any differences in the structures that lie behind teaching and learning will have marked consequences for the patterns of curricular coverage and achievement within the participating countries – and that in an important sense the differences that are seen are *intended* by the systems concerned in that they are the inevitable outcomes of cultures or structures of organization. Federalism or "local control" of schooling have consequences that override what might seem to be "national" expectations about what the formal outputs of schooling should or might be. In this section we will pursue this notion of intentional diversity and present a typology which we would expect to illuminate the pattern of achievement and the patterns of curricular coverage which we will see in the empirical data from the Study. Our goal in sketching this typology is not, of course, formal prediction but context-setting; to provide a vantage point that draws together some of the discussion in the previous sections into one structure.

It is clear that the *number of courses* available within a system and the form of curricular control found in practice within a school system create a variety of potentials for various forms of planned and unplanned diversity between the systems participating in SIMS. Table 3.11.1a sketches these structural forces and frames as they bear on Population A and Table 3.11.1b presents them as they bear on Population B. In each case the entries represent judgments about the central tendencies to be expected in each system and in some cases these entries may differ somewhat from those found in earlier cases. Thus, despite the presence of a variety of effective courses at the Population B level in Finland, we have coded that system both in Table 3.8.2 and Table 3.11.1b as having one course; and in the case of Scotland we have likewise discounted the existence of a second alternative course at the Population A level. In addition, in developing the classifications presented by this typology we have ignored history—so that, for example, in the case of Thailand we have neglected a recent history of considerable centralization. In the case of Canada (Ontario) we have likewise ignored the shifts that have occurred recently in the province's policies around central and local control and initiative.

The classification of systems into the cells of this typology can only be suggestive. But, at the level of suggestion, it does foreshadow the possibility of substantial variation in the patterns of content coverage and achievement that we might expect to see in later chapters of this volume and in the succeeding volumes of this Study. In the cases of both Population A and Population B we see all of the six cells containing entries and clear progressions from cases in which there are firm structural and framing

Table 3.11.1 *Planned Variation in Curricular Coverage: A Typology*

(a) Population A

	Location of Curriculum Control		
No. of courses taught at Pop A Level	Centralized	Mixed	Local
1	Belgium (Flemish) Belgium (French) France Hong Kong Ireland Japan	Canada (British Columbia) New Zealand Nigeria Scotland	Canada (Ontario) Israel Thailand
2	Sweden	Finland	
3+	Hungary	England and Wales Luxembourg Netherlands	United States

(b) Population B

	Location of Curriculum Control	
No. of courses at Pop B Level	Centralized	Local
1	Canada (British Columbia) Finland Israel Japan Sweden Thailand	
2	Belgium (Flemish) Belgium (French) Hungary Ireland New Zealand Scotland	
3+	England and Wales Hong Kong	Canada (Ontario) United States

constraints on coverage and achievement to cases in which there are few such constraints. At the Population A level, for example, France, Hong Kong, Japan, New Zealand and Nigeria are systems in which one course only is taught (at the level of intention at least) and the work of teachers and students is monitored closely by a central authority (via inspection, the backwash effects of external examinations, etc.); in the United States four or more courses are taught and what these courses mean in a particular context (state, local educational authority, school, classroom) is, at bottom, a local matter. We see the same kind of differences at the Population B level but here we have Finland, Japan, Nigeria, Sweden and Thailand falling into the cell we associate with maximal curricular coherence (i.e., one course only being taught) and the tightest control while Canada (Ontario) and the United States fall into the cell representing the greatest potential for diversity of both coverage and attainment. And, as we have emphasized, these are contexts in which the values which we might associate with these different framing structures are part and parcel of a larger body of sociocultural understanding about what institutions in these different nations should be like. Apologists for each context could (and do) defend the orders they know but in sum all one can say is that they are different orders.

In a very real sense this is the theme we have been developing in this chapter. The systems that make up the SIMS set are different in many respects. Comparison of one with the other is a complex enterprise and as we seek to achieve the control necessary for meaningful comparison there are many points of potential difference and similarity to be entertained. As we have seen, the organizational contexts of various systems differ and all of the factors and derived variables that one might canvass have clear potential for affecting both curricular practice and achievement.

3.12   Summary

This chapter has described the various SIMS populations and begun an exploration of the major differences between these populations. It will be the task of Chapter 4 to continue an investigation of the similarities and differences between these populations at the level of the *intended* curriculum. Chapter 5 will extend that exploration to the *implemented* curriculum. Thus, this chapter is but one part of the larger exploration of the dimensions of the comparability, and therefore the difference, between various populations included within the scope of SIMS. Such issues have an intrinsic significance but they are, of course, basic to any analysis and interpretation of the attained curricula found in the various participating systems.

What are the major conclusions of the analysis reported in this chapter? It would seem that at the Population A level there is a fair degree of structural similarity between the various target populations even though there are significant differences in the organization of curricula and courses and in the grade levels being investigated. It does seem reasonable, however, to suggest that a fair degree of homogeneity of the populations in the various systems has been achieved. This may or may not be the case, of course, for the *content* of the curricula of these populations.

There are more substantial problems at the Population B level. For example, there appear to be some potentially significant structural differences between national populations in the proportion of students in the target group, in the implicit tasks of schooling, and in the time available for teaching and learning. Needless to say, the issue of whether or not these differences affect the intended, implemented and attained curricula remains an open question. It is perhaps enough to say at this point that the analysis presented here might foreshadow some problems of interpreting some analyses at the Population B level. But, first, it needs to be seen how the differences between nations that have emerged in this chapter are reflected in the intended and implemented curricula of the participating systems.

4

The Content of the Intended Curriculum

The first three chapters have established the framework within which SIMS was undertaken. In this chapter we will consider some of the outcomes of the study: the mathematical content of the curriculum for the two target populations *as intended* by the participating systems.

As discussed in Chapter 2, SIMS *Working Paper I* was prepared as a guide for developing the international grids, one for Population A and one for Population B. The importance ratings in the grids, in turn, served as weights for determining the kinds of items (with respect to content and behavioral classification) and the number of items that were to comprise the international pool. Necessarily the grid and the pool represent a sort of international consensus as to the character of mathematics likely to be in the curriculum for the target population in each country. As expected, however, the fit of the item pool to the intended curriculum varied somewhat from system to system and from topic to topic.

The nature of this fit is described in terms of the commonality of subject matter across the intended curricula for all participating systems. We therefore can determine—of course within the limitations of our instrumentation—the extent to which school mathematics is similar across systems. Furthermore, and in some ways more interestingly, the unique characteristics of the curricula of the various systems can also be described.

4.1 Notation and Terminology

The mathematical content of the curriculum for a given population for a given educational system may be represented by Figure 4.1.1. Let the square encompass that body of content that is a candidate for inclusion in the curriculum for a target population. Then Regions II and III represent the mathematical content that is in the intended curriculum for the target population and Regions III and IV represent the mathematics that is in the implemented curriculum—the mathematics taught to the target population. The ellipse represents the content of the SIMS item pool.

79

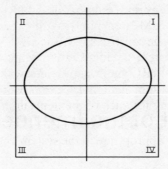

FIG 4.1.1   A Framework for Analyzing the Mathematics Curriculum

How well does the content of the SIMS pool match the intended curriculum for a given system? This question has two aspects.

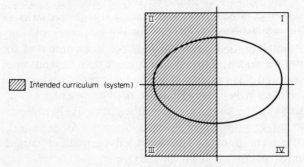

Intended curriculum (system)

FIG 4.1.2   Intended Curriculum System: The Portion of the
International Item Pool Appropriate for Target Populations

The shaded part of the ellipse represents that portion of the SIMS pool that was judged to be appropriate for the curriculum of a system. (That is, the shaded region representing the mathematics both intended to be taught and covered by the SIMS pool.) It is important to note that this does not encompass all of the SIMS pool nor all of any system's intended curriculum. It is also clear, from the manner in which the items of the pool were selected, that for a given system there would be items covering topics that are *not* part of its intended curriculum. The content covered by these items is represented by that portion of the ellipse outside the rectangle of Quadrants II and III.

Similarly, the implemented curriculum for a system (that mathematics reported to be taught to the target classes) may be represented by Quadrants III and IV in Figure 4.1.2. It therefore follows that the content represented by Quadrant IV is that mathematics reported to be taught but is not part of the intended curriculum. Quadrant III shows the content that is both intended and taught.

The SIMS methodology does not provide a direct measure of the regions outside the ellipse (the SIMS item pool). Within the pool, however, various indices of coverage may be constructed. In the SIMS' analysis of the intended curriculum of each system an index was computed from the following information: each National Center was provided with the complete item pool for each target population; National Centers were asked to provide an "appropriateness" rating for each of the items in the pool using this rating scale.

2—the item is highly appropriate to the system's curriculum for the target population.

1—the item is acceptable.

0—the item is inappropriate.

The instruction to National Centers when providing appropriateness ratings stated:

This (appropriateness rating) refers to the appropriateness of the item *for students* in each country. If the item tests knowledge or skill taught up to or at the level of the target population and is likely to be moderately difficult or easy at the end of the school year and does not have a strong cultural bias, then it is likely to be appropriate.

Using this information, intended coverage indices were computed for each item, for each content area and for each system. As an illustration of the procedure and notation used in this volume, the intended coverage data for Arithmetic at Population A are presented here.*

Table 4.1.1   *Indices of Intended Coverage for Arithmetic (Population A)*

System (Symbol)	Arithmetic N Items = 62
Canada (Ontario) (CON)	1.00
Scotland (SCO)	1.00
United States (USA)	1.00
England/Wales (ENW)	.98
Canada (British Columbia) (CBC)	.97
Japan (JPN)	.95
Thailand (THA)	.95
Hungary (HUN)	.93
Ireland (IRE)	.93
Luxembourg (LUX)	.93
New Zealand (NZE)	.93
Hong Kong (HKO)	.91
Israel (ISR)	.91
Belgium (Flemish) (BFL)	.90
Belgium (French) (BFR)	.90
Netherlands (NTH)	.89
Swaziland (SWA)	.87
Sweden (SWE)	.87
France (FRA)	.87
Finland (FIN)	.80
Mean	.92
Median	.93

* For the Population A systems the appropriateness ratings on all items used in the longitudinal study are presented in Volume III (Burstein et al. 1989). See also p. x above.

It can be seen from Table 4.1.1 that for three systems (Canada (Ontario), Scotland and United States), all of the Arithmetic items were judged to be appropriate for Population A. On the other hand, in Finland, only 80 percent of the items were judged to be appropriate. The mean coverage was .92 and the median was .93. Overall, then, the intended coverage of the item pool for Arithmetic was high.

These data can be displayed succinctly using a stem-and-leaf table.*

Table 4.1.2   *Stem and Leaf Diagrams for Intended Coverage Data of Table 4.1.1*

Stem	Leaf
10	000
9	001133335578
8	07770
7	
6	

Alternatively, using the three letter abbreviation code for each system, the indices may be displayed as follows:

Table 4.1.3   *Modified Stem and Leaf Table for Data of Table 4.1.2*

Stem	Leaf
10	CON SCO USA
9	BFR BFL ISR HKO NZE LUX IRE HUN THA JPN CBC ENW
8	FIN FRA SWE SWA NTH
7	
6	
5	
4	

It can again be seen that the set of arithmetic items was seen as highly appropriate for the participating systems. The systems for which coverage was high are Canada (Ontario), Scotland and the United States, with 100 percent of the items appropriate. Finland has lowest index of appropriateness.

---

* This method affords a method for recording data compactly, yet displaying overall patterns in the data. In this sense, the stem and leaf is like the more familiar frequency distribution. The added benefit of the stem-and-leaf is that, unlike the frequency distribution, the original data can be reconstructed from the display.

Hence, for example, the data 100, 93, 90, and 87 are displayed in a stem-and-leaf table as:

Stem	Leaf
10	0
9	03
8	7

4.2   Intended Content Coverage: Population A

Table 4.2.1 summarizes the indices of intended content coverage for those twenty systems for which appropriateness ratings were available. Intended coverage for Arithmetic and Measurement is very high for all systems, on average, with mean indices of .92 and .91 for Arithmetic and Measurement, respectively. Algebra is next in order of intended coverage with a mean index of .83. Geometry and Statistics have the lowest intended coverage, with mean indices of .64 and .69, respectively.

The structure of Table 4.2.1 as a modified stem-and-leaf plot also exhibits striking patterns of variation between systems from Arithmetic (on the left) to Statistics (on the right). That is to say, variation in Arithmetic is the smallest, with Algebra and Measurement having only slightly more between system variation in intended coverage. However, there is great variation in Geometry and Statistics; in Statistics, four systems (New Zealand, Scotland, Sweden and the United States) reported all of the items as appropriate while Israel reported none of the items as appropriate.

Table 4.2.1   *Population A: Indices of Intended Curricular Coverage*

	Arithmetic 000	Algebra 100	Measurement 400	Geometry 200	Statistics 300
10	CON SCO USA	IRE NZE	CON ENW HUN ISR JPN NZE SCO THA USA		NZE SCO SWE USA
9	BFL BFR HKO ISR HUN IRE LUX NZE JPN THA CBC ENW	FRA HUN SCO BFL BFR JPN	FIN HKO IRE CBC SWE	JPN NZE SCO	CBC
8	FIN FRA SWA SWE NTH	CBC ENW NTH	NTH SWA	ENW CON HUN	ENW FIN HKO HUN CON SWA
7		ISR THA SWE LUX FIN HKO SWA	LUX	FIN ISR THA NTH	JPN THA
6		USA CON	FRA BFL BFR	SWA HKO	NTH IRE
5			SWE USA		
4			CBC FRA		
3			IRE		
2				BFL BFR LUX	LUX FRA
1					BFL BFR
0					ISR
N. Items	62	42	26	51	18
Mean	.92	.83	.91	.64	.69

Figures 4.2.1 to 4.2.6 portray the general patterns of intended coverage in the curriculum of each system, based on the appropriateness ratings. All of these "intended coverage tables" have been constructed in a similar manner.

Consider, for example, Figure 4.2.1. The columns of the table, representing content areas, are ranked in descending order from left to right. That is, Arithmetic is on the extreme left, since it has, on average, the highest index of intended content coverage (Mean = .92) for the 20 systems. Another way of saying this is that of all the items in the international pool, those dealing with Arithmetic had, on average, the best fit to the intended

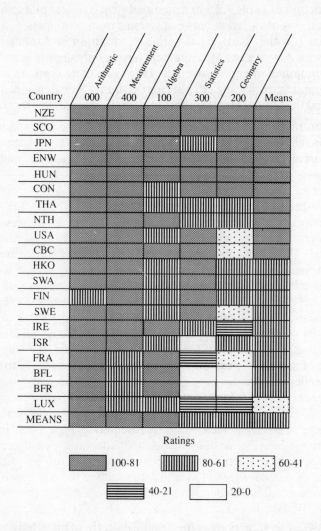

Country	Arithmetic 000	Measurement 400	Algebra 100	Statistics 300	Geometry 200	Means
NZE						
SCO						
JPN						
ENW						
HUN						
CON						
THA						
NTH						
USA						
CBC						
HKO						
SWA						
FIN						
SWE						
IRE						
ISR						
FRA						
BFL						
BFR						
LUX						
MEANS						

Ratings

100-81   80-61   60-41   40-21   20-0

Fig 4.2.1   Intended Coverage for Content Totals (Population A)

curricula of the systems. The next highest index, on average, was for Measurement, with a mean index of .91. The lowest mean intended coverage was for Geometry (.64).

The rows present the data for the systems, which are ranked from top to bottom on the basis of their mean index of coverage across the five content areas. The means are weighted by the number of items in each content area. We see that Scotland, New Zealand and Japan have the highest index of intended coverage and that Flemish and French Belgium and Luxembourg have the lowest. In other words, on average, the international pool of items fits the curricula of the former systems best and the latter systems least well. The shading indicates general levels of coverage. Darker shading represents higher indices. (See the key at the bottom of the figure.)

Two kinds of information can be obtained from these figures. First, the general darkness of the shading gives an indication of overall intended content coverage. Thus, for Arithmetic and Measurement, intended content coverage is high (the columns for these content areas are relatively dark). Alternatively, intended coverage for Statistics is low (shading of this column is relatively light). As already noted, intended coverage of the five content areas for Scotland, New Zealand, Japan, Hungary, and England and Wales is high.

Figures 4.2.2 to 4.2.6 provide intended coverage data at the content level. For Measurement, the table is overall very darkly shaded (Figure 4.2.4). By far the majority of systems have high intended coverage of Measurement. The exceptions are Belgium (Flemish and French), France, Luxembourg and Israel. These systems appear in the lower rows of the figure. Among topics, Approximation and Area and Volume receive fairly high coverage, with over 90 percent (on average) of the items rated acceptable or highly appropriate. Intended coverage for the remaining two topics, Estimation and Standard Units, is almost as high, with an average index of about 90 percent. By contrast, Figure 4.2.6, for Statistics, has large "chunks" of light gray shading, indicating that for many systems (indeed, the majority) and for three of the five topics, few if any of the items were judged to be part of their intended curricula.

The second sort of information about intended coverage that is available from the figures has to do with the amount of variation in the data. Differences in coverage indices are represented by differences in shading. Therefore, a heavily mottled or spotted appearance will reflect large between-system or between-content area differences in indices. The smoother the appearance of the figure, the smaller the differences between systems or content in content coverage. The generally smooth appearance of Figure 4.2.4, for Measurement, has already been noted. Figure 4.2.2, for Arithmetic, is also smooth, indicating relatively little variation between systems or between topics for this content area. Geometry, by contrast, has a table (Figure 4.2.5) that is heavily mottled.

These intended coverage data can be used in several ways. They give a general impression as to the fit of the item pool to the curricula of the participating systems and help identify the sources of variation of this fit. The data also provide background information for looking at data on the amount of mathematics taught by the teacher (index of implemented coverage) and learned by the student (index of attainment). For example, one would expect the Geometry items, overall, to be difficult for students in Luxembourg and France, since little of this content is in the intended curricula of those systems. On the other hand, the Arithmetic items were judged at least acceptable by most systems. One would reasonably expect these items to be relatively easy, overall. Note the assumption, of course, of a strong relationship between the intended and implemented curriculum. This assumption is explored in the subsequent chapters.

The data on the intended curriculum are now considered on a topic by system basis. *Throughout the discussion at the topic level, it should be kept in mind that in some cases the number of items is very small. Caution should be exercised in making generalizations about the topic beyond the items involved.*

Arithmetic

Figure 4.2.2 is relatively dark, indicating that overall, the content of the items is a part of the intended curriculum for each system. The mean indices of intended coverage range from a low of .80 for Finland to a high of 1.00 for Canada (Ontario), Scotland and the United States. The matrix does, however, have something of a mottled appearance, pointing to topics within Arithmetic that receive light coverage in some systems and heavy coverage in others.

Coverage of the items on Powers and Exponents, Common Fractions and Decimal Fractions is uniformly very high among all systems. The items on Number Theory, Ratio, Proportion and Percent, and Natural and Whole

10	CON SCO USA
9	BFL BFR HKO ISR HUN IRE LUX NZE JPN THA CBC ENW
8	FIN FRA SWA SWE NTH
7	
6	
5	
4	
3	
2	
1	
0	

Intended Coverage for Arithmetic (N. Items = 62)

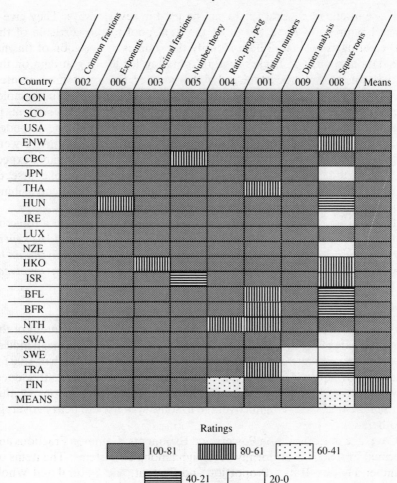

FIG 4.2.2   Intended Coverage for Arithmetic (Population A)

numbers also tend to be covered fully in all systems. Square Root was the topic in Arithmetic covered least well in the systems overall; the intended coverage of this topic is only .53, with Ireland, Japan, New Zealand, Swaziland, Sweden, and Finland reporting the Square Root items as "not appropriate" for Population A students. Low coverage of this topic is also reported for Hungary, Belgium (Flemish and French) and France.

Algebra

The predominance of dark gray regions in Figure 4.2.3 indicates generally high intended coverage of Algebra for all systems. On average, 83 percent of the items were judged acceptable or appropriate by the systems.

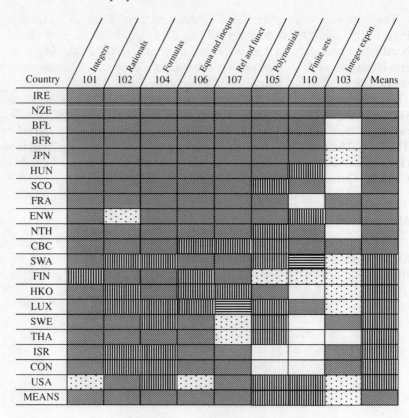

Ratings

| | 100-81 | | 80-61 | | 60-41 |
| | 40-21 | | 20-0 |

FIG 4.2.3   Intended Coverage for Algebra (Population A)

10	IRE NZE
9	FRA HUN SCO BFL BFR JPN
8	CBC ENW NTH
7	CON ISR THA SWE LUX FIN HKO SWA
6	USA
5	
4	
3	
2	
1	
0	

Intended Coverage for Algebra (N. items = 42)

The following topics, identified by the dark vertical bands on the left of Figure 4.2.3, had the highest overall intended coverage: Integers, Rational Numbers, Formulas and Algebraic Expression, Equations and Inequations, and Relations and Functions. The right side of Figure 4.2.3 has a much more spotted or mottled appearance. Intended coverage of Exponents and Set Theory is, on average, much lower than the preceding topics, and between-system variation in intended coverage is rather high. The items on Sets were rated as "not appropriate" in France, Hong Kong, Sweden, Israel and Thailand. The topic of Exponents was rated as "not appropriate" in Belgium (French and Flemish), Hungary, Scotland, The Netherlands, Thailand and Canada (Ontario).

Measurement

Figure 4.2.4 is predominantly dark gray and smooth, indicating overall high intended coverage of Measurement. The solid band across the top of the figure indicates that for most systems all Measurement items were part of

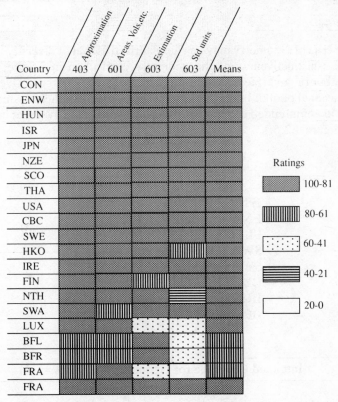

FIG 4.2.4   Intended Coverage for Measurement (Population A)

the intended curriculum. Coverage for Belgium (Flemish and French) and France is less heavy, but even for these systems the median index of coverage for all items is greater than .65.

10	CON ENW HUN ISR JPN NZE SCO THA USA
9	FIN HKO IRE CBC SWE
8	NTH SWA
7	LUX
6	FRA BFL BFR
5	
4	
3	
2	
1	
0	

Intended Coverage for Measurement (N. items = 26)

Geometry

The case of Geometry may be characterized as that of diversity. It is here that the distinctiveness of the geometry curriculum in France, Ireland, Luxembourg and Flemish and French Belgium emerges. Overall, the international pool of items does not fit the curricula of these systems, since the index of intended coverage for these systems is, on average, in the 20s. (See Figure 4.2.5.)

10	
9	JPN NZE SCO
8	ENW CON HUN
7	FIN ISR THA NTH
6	SWA HKO
5	SWE USA
4	CBC FRA
3	IRE
2	BFL BFR LUX
1	
0	

Intended Coverage for Geometry (N. Items = 51)

The dark gray regions identify a "common core" of geometry topics for all systems except French and Flemish Belgium. It should be noted that even within this common core, French and Flemish Belgium, Luxembourg, France and Ireland intend light coverage of the following topics: Classification of plane figures; Coordinates; Simple deduction. For the remaining topics, the indices of intended coverage have a bimodal distribution. There is a cluster of systems rating all of the items acceptable or appropriate and another cluster rating none of the topics acceptable. The topics dealt with are Similarity of Plane Figures, Informal Transformations, Spatial Visualization and Representation and, to a lesser extent, Pythagorean Triangles. Those systems tending to exclude these topics are French and Flemish Belgium, The Netherlands, France and Luxembourg, and, occasionally, Finland and Hong Kong.

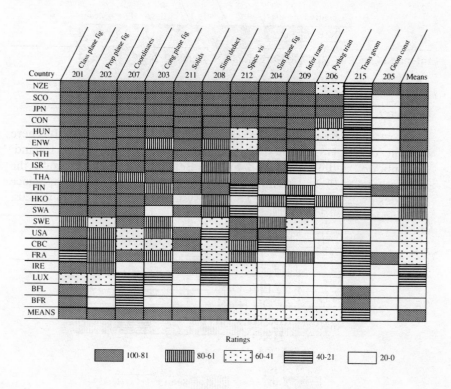

FIG 4.2.5   Intended Coverage for Geometry (Population A)

Statistics

This content area devotes only one item to Probability. The remainder deal with Organizing and Interpreting Data. The dark gray band across the top of the matrix indicates that for about one-third of the systems, the topics are covered rather heavily. Coverage in the remainder of the systems is light or nonexistent for most or all of the topics.

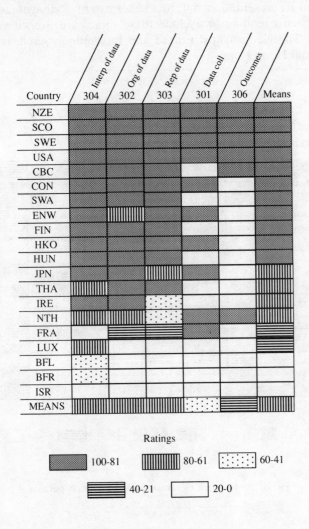

FIG 4.2.6   Intended Coverage for Statistics (Population A)

10	NZE SCO SWE USA
9	CBC
8	ENW FIN HKO HUN CON SWA
7	JPN THA
6	NTH IRE
5	
4	
3	
2	LUX FRA
1	BFL BFR
0	ISR

Intended Coverage for Statistics (N. Items = 18)

4.3   Curricular Clusters: Population A

It has been noted that coverage of Arithmetic and Measurement is high. This fact can be portrayed by plotting the intended coverage data for Arithmetic and Measurement as in Figure 4.3.1. The rather tight cluster of systems portrays coverage of these topics as being, in general, .80 or above. The exceptions are Measurement for France and Belgium (Flemish and French) (mean coverage of about .65).

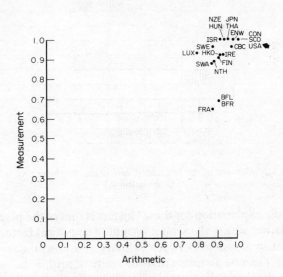

FIG 4.3.1   Content of the Intended Curriculum: Arithmetic vs. Measurement
(Population A)

By way of contrast, when Arithmetic is plotted against Algebra, two clusters emerge, "High Algebra" and "Low Algebra."

High Algebra	Low Algebra
Ireland	Swaziland
New Zealand	Finland
Belgium (Flemish)	Hong Kong
Belgium (French)	Luxembourg
Japan	Sweden
Hungary	Thailand
Scotland	Israel
France	Canada (Ontario)
England and Wales	United States
Canada (British Columbia)	
Netherlands	

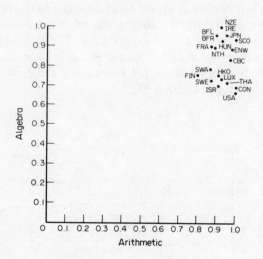

Fig 4.3.2   Content of the Intended Curriculum: Arithmetic vs. Algebra
(Population A)

One possible explanation for these clusters is that of the placement of the target population within the school system. The critical factor may be that division point in a system between primary and postprimary (secondary) education. If it can be assumed that the study of arithmetic is typically the province of primary education, then one could expect that the further along the target population is into postprimary education, the more likely it would be that algebra is a significant part of the curriculum.

Table 4.3.1   *Cluster Membership*

	High Algebra/High Arithmetic	Low Algebra/High Arithmetic
3	ENW FRA	
2	BFR BFL NET SCO	LUX SWA THA
1	CBC IRE JPN NZE	FIN HKO SWE
−1	HUN	ISR CON USA

Table 4.3.1 is displayed in such a way as to explore the possible relationship between importance of algebra and placement of Population A along the primary-postprimary divide. The data do support, to some extent, the hypothesized relationship. The exceptions are the "Low Algebra" systems in which Population A is well beyond the primary-postprimary divide, i.e., Luxembourg, Swaziland and Thailand. For most systems, year 1 is a transitional year between primary and secondary schooling.

The Bourbaki Tradition Another cluster of systems may be identified by looking at the overall match of the SIMS pool to their curricula, as indicated in Figure 4.3.2. Five systems, Ireland, France, Belgium (Flemish), Belgium (French) and Luxembourg have the lowest intended coverage ratings. The five systems have a distinctive curricular tradition due in large part to the influence of the Bourbaki Group in Europe, and especially in France and Belgium, in the 1960s.

Servais (1975) in reviewing curricular development in Europe captures the spirit of reform of the period:

> In *New Thinking in School Mathematics* we find already the contents of a unified mathematics syllabus: "The theory of vectors, linear algebra and parts of geometry (synthetic or analytic) are essentially three different languages for describing the same mathematical facts; for plane geometry, the theory of complex numbers provides a fourth interpretation . . . At the basis of all of this work is the arithmetic of real numbers."
> The rise of modern symbolism both in the introduction of sets and logic is advocated together with changes in types of graphs. Geometry must undergo a deep modification, and Professor J. Dieudonne with his slogan "Euclid must go" presented an outline of a program in which geometry was based entirely on a vector space of two dimensions.
> A new topic that appeared for the first time was probability and statistics including the normal distribution and statistical inference. (page 40)

This suggests a picture of a dramatically new curriculum, one that was perceived to appropriately reflect then-current developments in mathematics. As our findings suggest, however, the distinctive program of these reformers remained within the borders of only a few systems.

4.4   Intended Content Coverage: Population B

The same methodology is used here as was used in the discussion of Population A. For each content area and topic, an index of Intended Content Coverage was constructed that gives the proportion of items in the international pool, based on that content area or topic, that was judged to be a part of the intended curriculum for a particular system. Stem-and-leaf tables are also used to summarize the findings.

A summary of the indices of intended content coverage is given in Table 4.4.1, in a modified stem-and-leaf format. The table provides an overview of the content of the intended curriculum of the various systems *from the perspective of the international pool of items*. Note, for example, that the systems are relatively homogeneous with respect to the *Algebra* items. Most of the items are intended to be taught, and the range of C(I) is from 80 percent to 100 percent. With respect to *Elementary Functions and Calculus*, intended coverage is similarly high, with the exception of Thailand and Canada (British Columbia), whose indices are well below those of the other systems (in the 60 percent and 30 percent ranges, respectively). The mean for *Elementary Functions and Calculus* for all systems is .91, but when Thailand and Canada (British Columbia) are excluded this mean rises to .96. The indices for *Sets* and *Relations* form a bimodal distribution. About one-half of the systems intend to cover all the items. The remaining systems found between 70 percent and 80 percent of the items acceptable or appropriate to their curricula.

There is considerable variation in the intended coverage for *Geometry* with some systems (Canada (Ontario), and Japan) rating all the content appropriate and two systems (Thailand and Canada (British Columbia)) finding only about 50 percent of the items appropriate. Even larger ranges in the indices of intended coverage are found for *Probability and Statistics* and for *Finite Mathematics*.

Content by System Matrices

Figure 4.4.1 presents content area-by-system intended coverage for Population B. The columns represent the content areas, ranked from left to right according to decreasing order of the mean indices of intended content coverage. Again, the relative degree of coverage is indicated by the darkness of the shading. The left side of the matrix is darker, over all, than the right side, since the topics on the left are more intensively covered in the curricula of the systems than those on the right.

The rows of the matrix contain the systems, ranked from top to bottom in terms of their average index of intended content coverage for all content areas. Canada (Ontario) is toward the top of the matrix, characterized by a dark gray strip, since the pool of items tended overall to fit that system's curriculum rather well. Canada (British Columbia), on the other hand, is located toward the bottom of the matrix, and is represented by a light gray region. This indicates a rather poor fit of the items to that system's curricula.

Table 4.4.1.   *Population B: Indices of Intended Curricular Coverage*

	Sets and Relations	Algebra	Number Systems	Elementary Functions and Calculus	Geometry	Probability and Statistics	Finite Mathematics
	400	700	600	100	200	300	500
10	BFL BFR CBC CON FIN FRA HKO ISR LUX NZE SCO THA	BFL BFR CON ENW HKO IRE NZE SCO THA	ENW IRE ISR	NZE		CON ENW FIN FRA IRE JPN NZE	CON FIN FRA HKO IRE JPN LUX NZE SCO
9		FRA HUN JPN LUX SWE FIN ISR	BFL BFR FIN FRA HKO JPN LUX NZE	FRA IRE LUX SCO BFL BFR FIN HUN ENW HKO ISR JPN USA	HKO		
8	IRE JPN USA	CBC USA	CBC SWE USA	CON SWE	CON JPN IRE FRA LUX	HKO ISR LUX SWE THA USA	
7	ENW HUN SWE		THA CON SCO		FIN HUN ENW NZE		BFL BFR ENW SWE
6			HUN		ISR SCO THA SWE BFL BFR USA		
5				THA	CBC THA		USA
4						SCO	
3				CBC			
2						BFL BFR HUN	CBC
1						CBC	
0							HUN ISR THA
N Items	7	25	19	46	28	7	4
Mean	.93	.95	.89	.89	.73	.73	.71

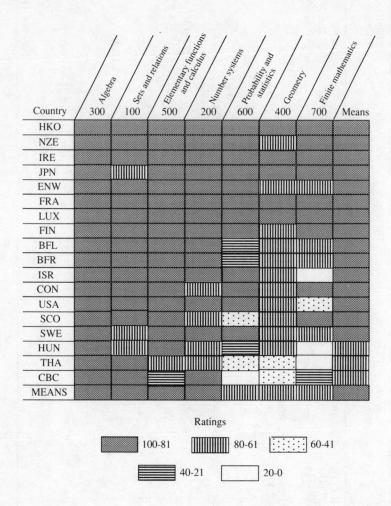

Fig 4.4.1   Intended Coverage for Content Totals (Population B)

Sets and Relations (100)

For this content area, no items are in the pool for the topics of Set Notation, Relations or Infinite Sets. For the remaining two topics intended coverage is widespread. Only England and Wales, Hungary, Ireland, Japan, Sweden and the United States have mean intended coverage indices for this content area of less than 1.00.

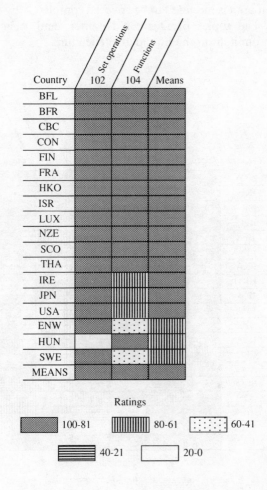

FIG 4.4.2   Intended Coverage for Sets and Relations (Population B)

10	BFL BFR CBC CON FIN FRA HKO ISR LUX NZE SCO THA
9	
8	IRE JPN USA
7	ENW HUN SWE

Intended Coverage for Sets and Relations (N. Items = 7)

Number Systems

This content area is intended to be covered completely by the majority of the systems. The topics of Decimal Numbers and Real Numbers are particularly prominent in the intended curriculum.

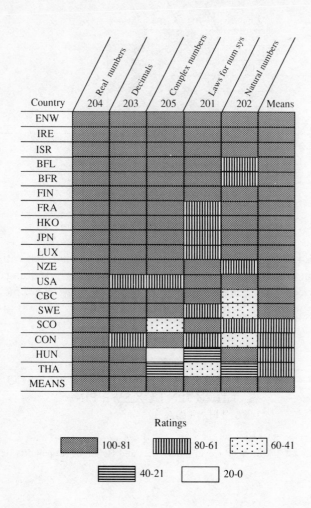

Ratings

100-81   80-61   60-41

40-21   20-0

FIG 4.4.3   Intended Coverage for Number Systems (Population B)

10	ENW IRE ISR
9	BFL BFR FIN FRA HKO JPN LUX NZE
8	CBC SWE USA
7	CON SCO
6	HUN THA
5	
4	
3	
2	
1	
0	

Intended Coverage for Number Systems (N. Items = 19)

Algebra

Figure 4.4.3 is relatively dark, and uniformly shaded, indicating a high and uniform intended coverage of the Algebra items. The striking exception is for Matrices, (at least for the one item representing this topic) which is reported not to be appropriate for the curricula of Finland, Israel and Hungary.

10	BFL BFR CON ENW HKO IRE NZE SCO
9	FRA HUN JPN LUX SWE FIN ISR
8	CBC THA USA
7	
6	
5	
4	
3	
2	
1	
0	

Intended Coverage for Algebra (N. Items = 25)

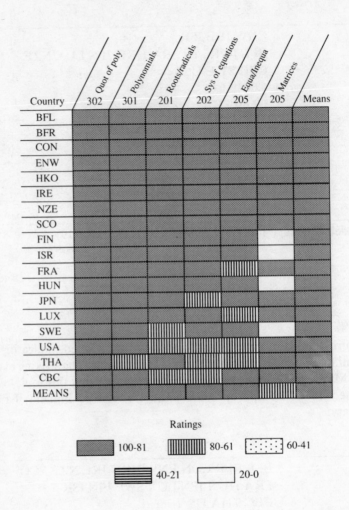

FIG 4.4.4   Intended Coverage for Algebra (Population B)

Elementary Functions and Calculus

The general dark and uniform appearance of this figure indicates a typically high intended coverage of this content area. The median index of Intended Coverage is .98. The exceptions are Thailand and Canada (British Columbia), whose intended coverage indices are .63 and .39, respectively. These two systems do deal with Elementary Functions, Properties of Functions, and Limits and Continuity but do not move on to Integration and Differentiation.

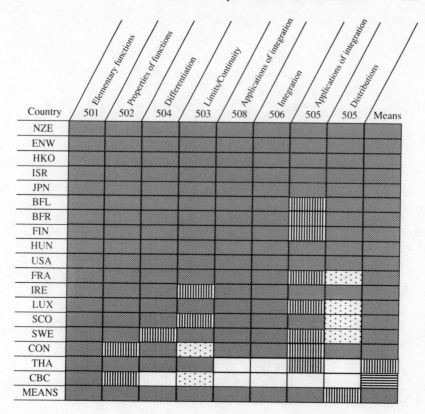

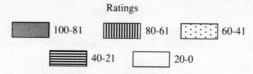

FIG 4.4.5   Intended Coverage for Elementary Functions and Calculus
(Population B)

10	NZE
9	FRA IRE LUX SCO BFL BFR FIN HUN USA ENW HKO ISR JPN
8	CON SWE
7	
6	THA
5	
4	
3	CBC

Intended Coverage for Elementary Functions and Calculus (N. Items = 46)

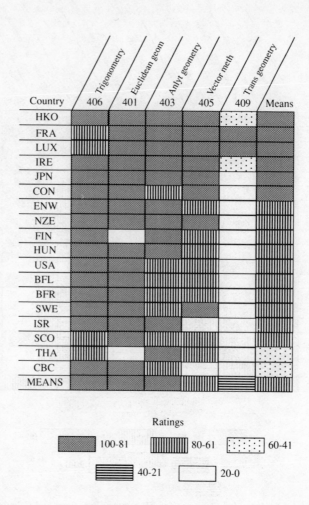

FIG 4.4.6   Intended Coverage for Geometry (Population B)

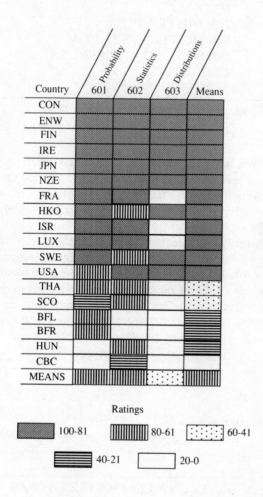

Ratings

100-81   80-61   60-41

40-21   20-0

FIG 4.4.7   Intended Coverage for Probability and Statistics (Population B)

Geometry

There are no items in the pool for Affine and Projective Geometry, Three-dimensional Coordinate Geometry, Finite Geometrics or Elements of Topology. Since there is only one item on Euclidean Geometry, this topic has been removed from the figure, but is included in the presentation below.

10	
9	HKO
8	JPN IRE FRA LUX
7	FIN HUN ENW NZE CON
6	ISR SCO SWE BFL BFR USA
5	CBC THA
4	

Intended Coverage for Geometry (N. Items = 28)

This content area has great variation in intended coverage, both between-systems and between-topics, as indicated by the mottled appearance of the figure. Trigonometry is the exception in that it has a mean C (I) of .90. Analytic Geometry is almost as high, with a mean index of intended coverage of .84.

Intended coverage of Vector Methods differs considerably between systems. Six systems include all of the items: Belgium (Flemish and French), Canada (Ontario), Japan, Hong Kong, France and the United States, while Israel and Canada (British Columbia) include only one of the items in their intended curriculum. Transformational Geometry, as in Population A, exhibits the most variation in coverage between systems. Belgium and France reported all items to be covered. The majority of the systems reported none of the items to be acceptable.

Probability and Statistics

10	CON ENW FIN FRA IRE JPN NZE
9	
8	HKO ISR LUX SWE USA
7	
6	
5	THA
4	SCO
3	
2	BFL BFR HUN
1	CBC
0	

Intended Coverage for Probability and Statistics (N. Items = 7)

The systems vary greatly in coverage of this topic. (It should be noted that there were no items in the pool on either Statistical Inferences or Bivariate Statistics. Since there is only one item on Distributions, this topic is deleted from the figure.)

Finite Mathematics

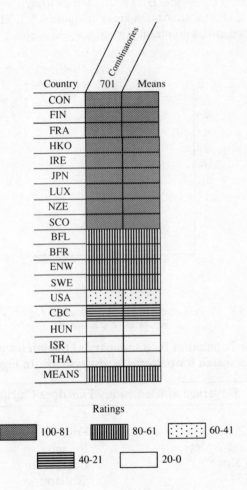

FIG 4.4.8 Intended Coverage for Finite Mathematics (Population B)

This content area is represented only by Combinatorics. Three of the sytems reported not including the topic in their curricula (Hungary, Israel and Thailand).

4.5   Content Clusters: Population B

As was done for Population A, indices of coverage for various content areas may be formed for Population B as well. It has been noted that for most systems, intended coverage of Algebra and of Sets and Relations is high overall for Population B. This is demonstrated by the plots of the coverage indices for these content areas in Figure 4.5.1. This subject matter, therefore, represents a common core of content across the systems.

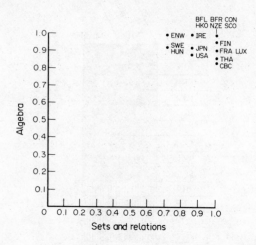

FIG 4.5.1   Content of the Intended Curriculum: Algebra vs. Sets and Relations (Population B)

Here, as for Population A, a summary of the intended coverage data is facilitated by a search for patterns across systems. In Figure 4.5.2 a plot is

Coverage of Elementary Functions/Calculus	
High	*Low*
New Zealand	France
England/Wales	Ireland
Hong Kong	Luxembourg
Israel	Scotland
Japan	Sweden
Belgium (Flemish and French)	Canada (Ontario)
Finland	Thailand
Hungary	Canada (British Columbia)
United States	

given of the intended coverage for Algebra against Elementary Functions and Calculus. Overall, coverage is high (greater than 80 percent) on Algebra for all systems. However, coverage is less high for Elementary Functions and Calculus. Hence, two clusters of systems may be identified.

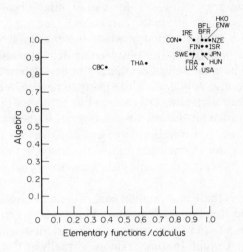

FIG 4.5.2   Content of the Intended Curriculum: Algebra vs. Elementary
Functions/Calculus (Population B)

The clustering for Elementary Functions and Calculus is not at all clear cut. France, for example, appears in the lower cluster because of relatively less emphasis on Applications of the Derivative and on Techniques of Integration than is reported for systems in the high group. Apparently, Ireland is slightly below the mean because of less content dealing with Limits and Continuity.

It seems safe to say most systems have taught Measurement, Arithmetic, Algebra, Geometry by the end of the Population A year, and have taught these topics together with Calculus to the students in Population B. The emphasis among the topics may have differed, or an individual topic (notably Calculus in some systems) may not have been emphasized. Such variations are, of course, of major interest in themselves, and may form a significant part of the analysis. However, the coverage indices for these topics may not be the only, or the best, indicators of a system's philosophy of mathematics education or its response to notable movements in the field. In this respect, the distinctive nature of a system's curriculum may be reflected more by the *other* topics that it teaches that are not represented in the SIMS grid. That is to say, by its attention not only to set theory and abstract algebra, but also to areas such as probability and statistics, logic, and computer science.

4.6   Summary

An analysis of the content of the intended curriculum yields patterns that reveal both uniformity and diversity. At the Population A level, Measurement and Arithmetic comprise a "common core" of mathematical content for all countries. However, for a substantial subset of the systems, Algebra is an important topic in the curriculum. One factor that appears to determine this importance is that of whether Population A is regarded as belonging to the primary or the postprimary school phase of a system's structure. There is some indication that when Population A is part of the *post*primary structure, Algebra is more likely to be included in the curriculum. There is great variety in Geometry content across the systems. Few systems include noteworthy amounts of statistics in their curricula. Those systems in the Bourbaki tradition have a distinctive curricular approach, characterized by a more formal and unified treatment of the subject matter, especially in Geometry but this approach seems to be confined to few SIMS systems.

For Population B there is a common core of content consisting of topics in Algebra, Sets and Relations, Elementary Functions and Number Systems. However, there is a rather sharp contrast in the extent to which the Calculus is a part of the curriculum. In some systems, virtually no Calculus is studied. In others, the subject is treated rather fully. A more detailed study of contextual factors leading to the study of Calculus in Population B is carried out in Chapter 6.

5
The Content of the Implemented Mathematics Curriculum*

5.1 Introduction

Chapter 1 emphasized the central place that the curriculum played in the design and conceptualization of SIMS. That chapter also sketched the overall organization of the ambiguous term "curriculum" that the study offered: the "curriculum" may be seen as a structure that is *intended* to control the scope of what should be done in schools; it may be seen as the body of content and practices that are in fact being *implemented* in schools; it may be seen as something that is realized or achieved in the understandings of students. In SIMS, these different aspects of the curriculum have been termed the *intended curriculum*, the *implemented curriculum* and the *attained curriculum*. One important task of SIMS was the exploration of the relationships between these discrete but at the same time related aspects of the curriculum.

Earlier IEA studies have explored aspects of these questions and, as a result, have served to emphasize the centrality of the implemented curriculum in influencing achievement. Within these studies the implemented curriculum was termed "opportunity-to-learn" (OTL) and a measure of OTL was secured by asking teachers in the schools included in each study's sample to rate the appropriateness of the items on the test for their students in light of the curriculum of that school. Where the influence of OTL on achievement was investigated, it was usually found that there were clear relationships between the extent of coverage of items and achievement. (See, for example, Comber and Keeves, 1972; Purves, 1987.)

The design of SIMS drew heavily on this tradition within the IEA studies, both in its emphasis on the curriculum as a variable which might influence achievement and in its operationalization of an approach to the measurement of opportunity-to-learn and, by implication, of the implemented curriculum. Teachers in the sampled classes were asked to respond to an instrument (the *Teacher Opportunity-to-Learn Questionnaire*) in which they

* The data on which this chapter was based were provided by Richard G. Wolfe of the Ontario Institute for Studies in Education, Toronto, Ontario, Canada.

were asked the following questions about each item in the student test forms used in their classroom:

1. During this school year, did you teach or review the mathematics needed to answer the item correctly?
   a. Yes
   b. No
2. If, in this year, you did NOT teach or review the mathematics needed to answer this item correctly, was it because,
   a. It has been taught prior to this school year?
   b. It will be taught later (this year or later)?
   c. It is not on the school curriculum at all?
   d. For other reasons.

Sixteen systems at the Population A level* and 12 systems at the Population B level† included the *Teacher Opportunity-to-Learn* (OTL) *Questionnaire* in their testing batteries. Our task in this chapter will be to review the findings that emerged from this instrument at the level of the subject area or content domain. We will focus on responses to a composite variable derived from questions (1a) and (2a) above. Thus we will be describing the implemented curriculum in each participating system *up to the point of testing*. It should be noted that this composite measure does not define the curriculum of the target year alone but instead defines the background knowledge that each class's teacher presumed that his or her students brought to the testing situation. Such a measure is, of course, subject to the vagaries of each teacher's knowledge of the work done in prior years and, in the case of some systems at the Population B level, to the interpretation that individual teachers made about the relevance of work that was being covered in parallel classes in mathematics. (See Chapter 3 above.)

In the pages that follow we seek to describe the implemented curriculum experienced by students in the Population A and Population B years in the content domains used within SIMS, i.e., for Population A, Arithmetic, Algebra, Geometry, Statistics, and Measurement and for Population B, Sets and Relations, Number Systems, Algebra, Geometry, Elementary Functions and Calculus, Probability and Statistics, and Finite Mathematics. In doing this we not only seek to set a context for the interpretation of the achievement patterns that will be reported in Volume 2 but we also attempt a beginning description of the curriculum of the various systems as phenomena in their own right.

---

* Belgium (Flemish), Canada (British Columbia), Canada (Ontario), England and Wales, Finland, France, Hungary, Israel, Japan, Luxembourg, Netherlands, New Zealand, Nigeria, Swaziland, Sweden, Thailand and the United States.
† Belgium (Flemish), Canada (British Columbia), Canada (Ontario), England and Wales, Finland, Hungary, Israel, Japan, New Zealand, Sweden, Thailand and the United States.

5.2   Measurement of Content Coverage: The Validity of OTL

The first question that arises as we begin this task is that of the validity of the data derived from the *Teacher OTL Questionnaire*. Two approaches to the assessment of the outcome of this instrument were undertaken. One considered the relationship between the measures of appropriateness considered in Chapter 3 and teacher reports of coverage. The other explored the relationship between teacher coverage and achievement.

5.2.1   Intended Coverage and Implemented Coverage

The index of intended coverage discussed in Chapter 4 is a measure that closely parallels the index of implemented coverage to be discussed in this chapter. Both are based on assessments about whether items in the SIMS pool had been covered in a curriculum, but both the referents and the respondents differ. In the case of the appropriateness ratings the question was asked of each SIMS National Committee and was asked in the context of each system's formal curriculum. In other words, "appropriateness" refers to the match between an item and the *intended* curriculum of the system whereas coverage refers to the match between an item and the experience of the students in a participating classroom as this was understood by the teacher of that class.

As we begin the task of assessing the match between the intended and implemented curricula of the participating systems, it needs to be recalled that the appropriateness measure has several problems associated with it. In several of the participating systems there is no national curriculum in the sense of an authoritatively-prescribed or -recommended curriculum. In these cases, National Committees were required to make judgments about the appropriateness of items based on their assessments of school practice, reviews of commonly-used textbooks and the like. Judgments of the appropriateness of items to a curriculum that are based on such sources are obviously different from those based on official syllabi. In addition, where several courses were available to students either as part of a formal organization of the curriculum of a system or in school practice, both National Committees and the International Mathematics Committee faced a difficult problem of weighting the appropriateness of items in different courses to produce a single summary judgment. Several national committees reported that, in the hindsight of the data provided by the *Teacher OTL Questionnaire*, their solutions to these problems were unsatisfactory and that their judgments of appropriateness not valid.

Figure 5.2.1 plots the relationship between the appropriateness ratings across each Population A content domain and the related indices of implemented coverage derived from the *Teacher OTL Questionnaire*.

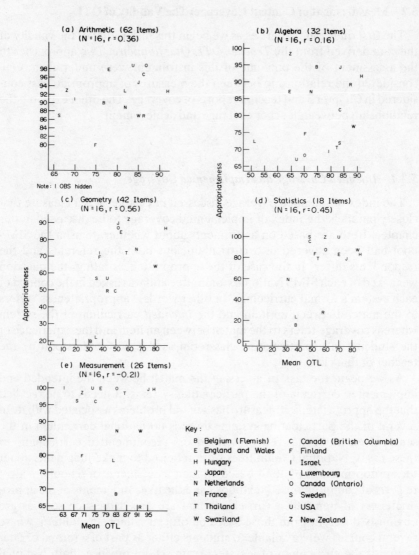

FIG 5.2.1   Population A: Appropriateness Ratings and Opportunity-to-Learn

Figure 5.2.2 reports the parallel findings for Population B. An inspection of Figures 5.2.1 and 5.2.2 suggests that there were substantial overall discrepancies between these two measures. This finding raises obvious problems as we think about the validity of either or both of these measures.

What is the major problem in the relationships as seen in these figures? Intentions run ahead of implementation in many systems at both population levels. In some cases this overambition on the part of national committees

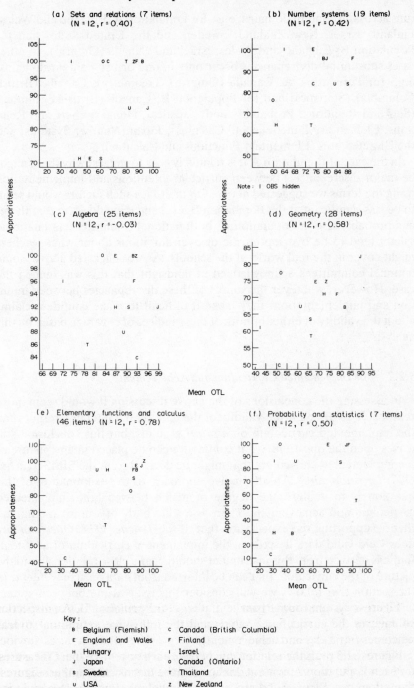

FIG 5.2.2   Population B: Appropriateness Ratings and Opportunity-to-Learn

runs across all content domains, e.g., for Population A, England and Wales, Finland, Israel, New Zealand, Sweden and the United States and for Population B, England and Wales, Israel and Canada (Ontario). In other cases substantial discrepancies occur only in one or two content domains, e.g., for Population A, Canada (Ontario) (Geometry), Canada (British Columbia) (Statistics) and for Population B, Canada (British Columbia) (Sets and Relations, Probability and Statistics), Hungary (Sets and Relations, Elementary Functions and Calculus), Japan (Number Systems) and the United States (Elementary Functions and Calculus).

In the case of Population A it is relatively easy to suggest why there might be major discrepancies between curricular intentions and implementation using the terms we suggested above. On their face such factors would seem to be less significant at the Population B level and it would seem that there, and probably also in Population A, both national committees and national syllabi tend to be over-optimistic or over-ambitious about what teachers might cover in the real world of the schools. As we suggested above, some national committees acknowledged in hindsight that this was indeed the case. However, whatever the source of these discrepancies between intention and implementation, they make it difficult to make confident claims about the validity of either or both of these indices of coverage based on this analysis.

5.2.2   *The Implemented Curriculum and Achievement*

In assessing the conclusions of the above discussion it would seem more reasonable to discount the validity of the data on the intended rather than the implemented curriculum on *a priori* grounds, but this conclusion still leaves open the question: What confidence can be placed in any picture of the implemented curriculum that might be derived from the SIMS *Teacher OTL Questionnaire*? An alternative approach to the examination of this question is to explore the degree of match between the implemented curriculum and achievement. There is a solid body of empirical data and theory supporting the expectation that, if the *Teacher OTL Questionnaire* does yield valid data describing the implemented curriculum of a system, findings derived from this instrument should bear a close relationship to the picture of the curriculum that can be inferred from data on achievement. In the section that follows we will consider this issue at the between-system level; cross-system comparison is, of course, the principal focus of both this volume on the curriculum analysis and the following volume on student outcomes (attitudes and achievement).

Figure 5.2.3 plots the relationship betwen each system's mean OTL rating for each Population A content domain and the mean achievement scores on those domains. Figure 5.2.4 reports the parallel relationships for the Population B content domains.

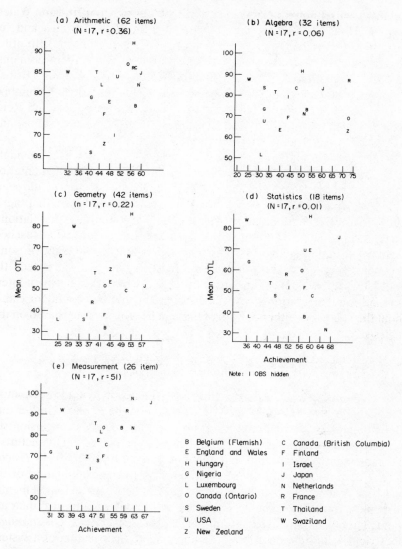

FIG 5.2.3   Population A: Opportunity-to-Learn and Achievement

Population A: Looked at overall, the plots of the relationship between teacher OTL ratings and achievement presented in Figure 5.2.3 do not suggest a firm and unequivocal correlation between the indices of coverage and achievement across the set of systems. However, a careful inspection of the plots indicates that there are systems with persistent and marked discrepancies between OTL and achievement, e.g., Nigeria (indicated on the plot by the character "G"), Swaziland (W), Hungary (H). If these

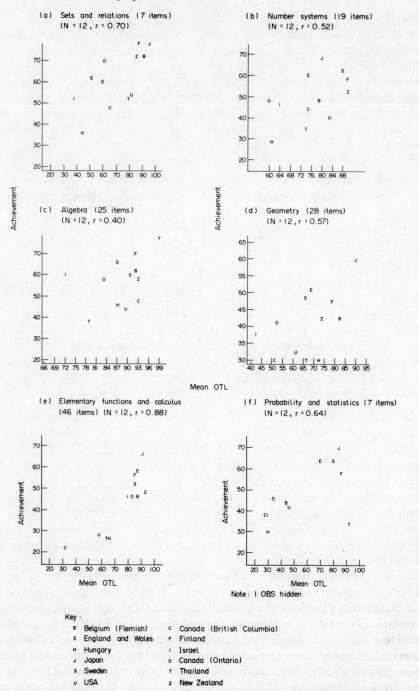

FIG 5.2.4   Population B: Opportunity-to-Learn and Achievement

systems are factored out as we inspect the plots, something approaching an overall linear relationship between OTL and achievement would seem to emerge although there are clear outliers in several of the content domains. Thus, in the content domain of Algebra, Sweden (S) and, in the domain of Measurement, the United States (U) would seem to have lower achievement than their OTL index would seem to foreshadow; in the domain of Algebra, Canada (Ontario) (O) and New Zealand (Z) and, in Statistics, Belgium (Flemish) (B) and the Netherlands (N) would seem to have higher achievement than their coverage would seem to foreshadow. However, within the limitations suggested by these observations, it would seem that overall the mean teacher OTL ratings do provide a reasonable predictor of between-system differences in achievement and have, therefore, some predictive validity at the system level. *It is not so clear that this is the case for the OTL ratings from Hungary, Nigeria and Swaziland and there must be questions to be raised about the validity of the coverage indices for these systems.*

Population B: The plots of the Population B between-system relationships of mean teacher OTL ratings and achievement are set out in Figure 5.2.4. (No plot is presented for the content domain of Finite Mathematics because of the small number of items in that domain (4 items). It also needs to be noted that there are comparatively few items in the domains of Sets and Relations (7 items) and Probability and Statistics (7 items)). The following discussion will focus on the findings for Number Systems, Algebra, Geometry and Elementary Calculus and Functions.

Looked at overall it would seem that the relationship between Population B coverage and achievement is tighter than in the case of the Population A subject areas in all domains except Number Systems. On closer inspection Canada (British Columbia) (C), Hungary (H), Thailand (T) and the United States (U) seem to emerge as persistent outliers—with higher reported coverage than their achievement would seem for foreshadow across several domains and Israel would seem to emerge as a clear outlier in the domains of Number Systems and Algebra. When these systems are factored out in an inspection of the plots something approaching a firm linear relationship across the remaining systems would seem to emerge across all areas—although, in considering this conclusion, it needs to be noted that the range of between-system coverage is quite restricted in the domains of Algebra and Elementary Functions and Calculus. It would seem that, as in the case of Population A, we can conclude that the aggregated teacher OTL ratings do provide a reasonable predictor of between-system differences in achievement and have therefore some predictive validity. However, *some questions about the validity of the OTL ratings from Hungary, Israel (Number Systems and Algebra), Thailand and the United States must remain.* This should be borne in mind in the course of the following discussion.

5.3   Dimensions of the Implemented Curriculum

As noted above, the data provided by the SIMS *Teacher OTL Questionnaire* not only have a different referent from the data provided by National Committees on the appropriateness of items, but also are different in character. Thus, while appropriateness ratings were reported at the system level and were based on one source, the data derived from the *OTL Questionnaire* reflects the ratings of all teachers in the national sample. *As a result the OTL data have the potential of reflecting the variety of teacher coverage within a system*—but at the same time, this variety poses problems for presentation and analysis.

In conceptualizing the possibilities that can emerge from the results of the *OTL Questionnaire*, SIMS drew heavily on the thinking underlying item-response analysis of *patterns of achievement* using Student-Problem (or S-P) charts (Cliff 1983; Harnisch & Linn 1981). As Harnisch and Linn (1981) point out in their discussion of such approaches to test analysis, a focus on the summary scores that can be derived from any test has serious limitations. Similar scores indicating the total number of correct responses to test items can be obtained in different ways and it is *patterns* of response that can be most useful and meaningful in identifying group differences in learning.

The S-P chart used for the analysis of such patterns of achievement is a matrix of zeros and ones that is doubly ordered; rows (students) are ordered from top to bottom in descending order of total number correct; columns (test items) are ordered from left to right in ascending order of difficulty. When both student achievement and item difficulty are perfectly ordered, a perfect Guttman scale results with the top row filled with all ones and the bottom row filled with all zeros. In practice, however, such a pattern rarely occurs and instead patterns like those seen in Figure 5.3.1(b) are found. In this figure, for example, the student in the fourth row from the top missed several of the easier items, i.e., items passed by many students, and passed several of the more difficult items. An analysis of S-P charts helps detect such lack of predictability in student performance and, to supplement visual analysis, several indices have been proposed to portray both student and item "predictability" (Cliff 1983).

In SIMS this approach to the analysis of achievement patterns was used to conceptualize the teacher OTL data. While the approach is identical to that associated with item-response analysis, the interpretation is different. The matrix represents items (rows) and classes (columns) that are then ordered from top to bottom and left to right in terms of extent of overall coverage. The resulting matrix offers a depiction of the extent of coverage by classes of the item domain, the variability of such coverage both across items and classes, and the predictability of coverage seen in terms of the typical classroom within a system. And, as is the case with an S-P chart, these dimensions of the chart can be summarized in terms of a set of indices

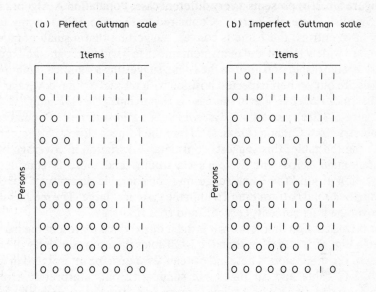

FIG 5.3.1   Perfect and Imperfect Guttman Scales

representing the *extent of coverage* of a total content domain, *variation* in coverage between classes within the system, and *diversity* of coverage seen in terms of deviation from the coverage pattern of the typical classroom. Coverage is represented by the mean (or median) percentage of items reported as covered within a system; variation is represented by the semi-interquartile range *(Q)*, and diversity can be represented by van der Flier's index of predictability *U'*, which measures mean between-class deviation from the modal pattern of coverage (van der Flier 1977, 1982).

Figure 5.3.2 illustrates how the results of the OTL Questionnaire can be portrayed in such an Item-Classroom (I-C) Table. Figure 5.3.2(a) presents the table on Arithmetic from Japan. The broken line (the I-line) running from the upper right corner of the table to the lower left is a marker for the number (proportion) of classes which were taught each item and represents the total number of classes taught that item. Diversity is reflected visually by the degree of "confusion" in the pattern of zeros and ones around this line.

In Figure 5.3.2(a) it is clear that most cells in the table have entries indicating that most items were taught to virtually all classes; the I-line is extremely concave, suggesting that there was little variation in the overall pattern of coverage within the system, and there is little apparent dispersion around the I-line. The median index of coverage derived from this table is 85 percent, the semi-interquartile range of the coverage scores *(Q)* is 2.0, and the mean *U'* is .03.

Figure 5.3.2(b) presents a very different case, Population A Algebra from the Swedish General Course.* Compared to Figure 5.3.2(a), many fewer cells have entries, the I-line is convex, suggesting that a small number of items are widely taught and many items much less widely taught, and there is considerable dispersion around the I-line, suggesting a significant number of classes are out of their expected position with respect to item coverage. The median index of coverage in this case is 28 percent, Q is 11.7 and U' is .15.

Figure 5.3.2(c) presents a case which is again different. Population A Geometry from Canada (Ontario). Here the I-line is close to a straight line bisecting the table; this suggests considerable variation in coverage of the total domain across the system's classrooms and there is considerable dispersion around the I-line suggesting considerable diversity of coverage within the broad pattern found within the system at large. The median index of coverage is 49 percent, Q is 16.7 and U' is .16.

In the discussion that follows it is the indices which represent the patterns seen in Figure 5.3.2 rather than the I-C Tables themselves which will be the focus of the discussion. These indices can, of course, be interpreted in terms of an I-C Table and, at times, we summarize our interpretation of the patterns seen in the set of systems in terms of patterns seen in such tables. In particular, we will refer at times to the three kinds of I-lines seen in Figure 5.3.2; concave and right-angled—see Figure 5.3.2(a) for Japan— implying substantial uniformity of coverage of the items to the left of the I-line; convex—see Figure 5.3.2(b) for Sweden—implying substantial coverage of a small set of items within the domain and much lower levels of coverage of the other items in the domain; and a straight line bisecting the table—see Figure 5.3.2(c) for Canada (Ontario)—implying considerable variation in coverage across a system's classrooms.

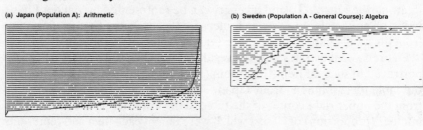

(a) Japan (Population A): Arithmetic

(b) Sweden (Population A - General Course): Algebra

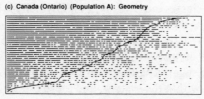

(c) Canada (Ontario) (Population A): Geometry

FIG 5.3.2   Some Illustrative I-C Tables

* This is one of the two courses that comprise Sweden's Population A.

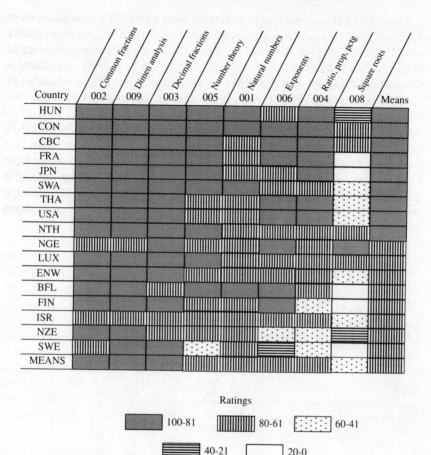

FIG 5.4.1   Population A: Implemented Coverage for Arithmetic

5.4   Population A: Patterns of Content Coverage

5.4.1   Arithmetic: Between-system Patterns of Coverage

Table 5.4.1 presents the weighted mean and median OTL indices, the first and third quartiles, the semi-interquartile ranges *(Q)* and the median van der Flier indices *(U')* for Arithmetic from each system.[*] Figure 5.4.1 presents a system-by-subtopic display of the implemented coverage of each subdomain within Arithmetic as indicated by mean teacher OTL scores. The shading on this figure indicates the general level of teacher coverage with darker shading representing higher OTL indices.

[*] Only mean OTL indices are available from Nigeria and Swaziland.

The overwhelming majority of systems have mean OTL indices in Arithmetic of over 70 percent with Canada (British Columbia), Canada (Ontario), France, Hungary, Japan, The Netherlands, Swaziland and the United States having indices above 80 percent. Figure 5.4.1 suggests that coverage of most subtopics within the SIMS content domain is, on the whole, high with the exception being the topic of Square Roots which has very low coverage reported from Belgium (Flemish), Finland, France, Japan and Sweden. Overall, the domain of Arithmetic as this is defined within SIMS, is broadly representative of the implemented curriculum in most systems.

However, it should be noted that Israel, New Zealand and Sweden have OTL indices below 70 percent. There are, in other words, exceptions to the generalization that Arithmetic as defined within the SIMS item pool is broadly representative of the implemented curriculum of the participating systems.

Table 5.4.1   *Population A: Coverage of Arithmetic*

	Mean (Weighted)	Q3	Md	Q1	Q	U'
Belgium (Flemish)	77	85	78	70	7.5	.09
Canada (British Columbia)	86	91	86	80	5.5	.15
Canada (Ontario)	87	94	89	80	7.0	.13
England and Wales	78	87	78	63	12.0	.18
Finland	75	80	74	67	6.5	.09
France	86	89	85	80	4.5	.12
Hungary	92	94	87	67	13.5	.10
Israel	70	80	59	44	18.0	.27
Japan	85	87	85	83	2.0	.02
Luxembourg	79	89	80	73	8.0	.17
Netherlands	82	86	81	73	6.5	.13
New Zealand	68	76	65	59	8.5	.13
Nigeria	79	NA	NA	NA	NA	NA
Swaziland	85	NA	NA	NA	NA	NA
Sweden	66	75	68	60	7.5	.09
Thailand	85	91	85	80	5.5	.10
United States	84	91	85	78	6.5	.11
Mean	80	80	79	69	8.0	.13
Median	82	87	81	70	7.0	.12

5.4.2   *Arithmetic: Within-system Patterns of Coverage*

Q is an index of the variation in coverage of the set of topics taught within a system and U' an index of the diversity of coverage found *within* the overall pattern of variation within a system. France, Hungary and Japan have the lowest Q's suggesting the least variation in coverage across the classrooms among the SIMS systems. England and Wales, and Israel have the highest variation and New Zealand and Luxembourg also have Q's which are above the midpoint of distribution of Q's. Japan has the lowest U' suggesting that this system also shows the least diversity in its pattern of coverage and Israel has the highest, suggesting the greatest diversity of coverage. Most of the systems appear, however, to be broadly similar in their level of diversity.

5.4.3   Measurement: Between-system Patterns of Coverage

Table 5.4.2 presents the indices of coverage for each system in the content domain of Measurement. Figure 5.4.2 presents a system by subtopic display of the coverage of each subdomain within the area of Measurement. As might be expected, the median OTL indices for Measurement are high overall (between system median 82 percent) and Finland (74 percent), Israel (54 percent), New Zealand (71 percent) and Sweden (71 percent) stand out as having much lower coverage than other systems. As would also be predicted, all subtopics within Measurement are widely covered.

Table 5.4.2   *Population A: Coverage of Measurement*

	Mean	Q3	Md	Q1	Q	U'
Belgium (Flemish)	83	96	83	75	10.5	.13
Canada (British Columbia)	75	92	84	55	18.5	.20
Canada (Ontario)	83	96	88	79	8.5	.16
England and Wales	79	92	79	65	13.5	.24
Finland	70	87	74	57	15.0	.15
France	92	96	92	83	6.5	.15
Hungary	97	100	96	79	10.5	.00
Israel	63	75	54	21	27.0	.33
Japan	95	100	96	92	4.0	.02
Luxembourg	82	86	83	78	4.5	.07
Netherlands	83	86	84	73	6.5	.19
New Zealand	70	88	71	54	17.0	.21
Nigeria	71	NA	NA	NA	NA	NA
Swaziland	92	NA	NA	NA	NA	NA
Sweden	68	80	71	55	12.5	.15
Thailand	86	91	88	75	8.0	.19
United States	74	92	79	63	14.5	.22
Mean	80	90	81	66	12.4	.16
Median	82	92	83	65	12.5	.16

5.4.4   Measurement: Within-system Patterns of Coverage

It is interesting to note that there is greater overall variation in the coverage of Measurement than was the case with Arithmetic. Within the overall pattern of variation, France ($Q = 6.3$), Japan ($Q = 4.2$), and Luxembourg ($Q = 4.7$) stand out as having the lowest variation and Israel ($Q = 27.1$) stands out as having the most variation in coverage. Figures 5.4.3(a) and 5.4.3(b) present the I-C tables for Japan and Israel and the

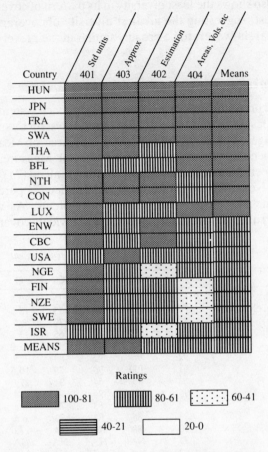

Ratings

100-81   80-61   60-41

40-21   20-0

FIG 5.4.2   Population A: Implemented Coverage for Measurement

differences in the underlying patterns of coverage in these two systems are clear. The more typical pattern is represented by the I-C table for Canada (British Columbia) (see Figure 5.4.3(c)) where the overall pattern is marked by a set of plateaus in the I-line and in the table for New Zealand, which is marked by a long gradual slope in the I-line (see Figure 5.4.3(d)).

It is difficult to interpret what such patterns, and the associated higher Qs might mean. Perhaps they reflect the reality that there are pupils at the Population A level in most systems who are having difficulties with elementary mathematical concepts and the consequent need for basic classes at these levels. However, it is not clear why this should be more clearly the case in Measurement than Arithmetic and it may be, as was suggested earlier, that this finding is, at bottom, a reflection of the varied understanding on the part of secondary teachers of the patterns of coverage found in the elementary grades. This latter interpretation might be supported by an inspection of the diversity indices which are also higher overall than the parallel indices for Arithmetic. Again, however, it is interesting to note that Japan shows the least diversity in its overall pattern of coverage ($U' = 00$) and Israel the highest ($U' = .33$). Within these extremes the United States ($U' = .22$), New Zealand ($U' = .21$) and Canada (British Columbia) ($U' = .20$) have the highest diversity indices and Luxembourg ($U' = .07$) the lowest.

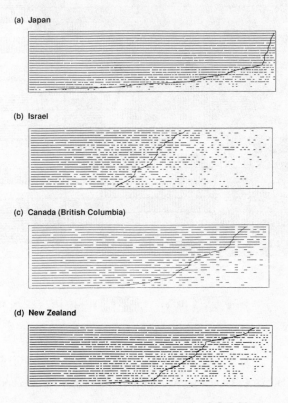

(a) Japan

(b) Israel

(c) Canada (British Columbia)

(d) New Zealand

FIG 5.4.3   Measurement: Illustrative I-C Tables

5.4.5   Algebra: Between-system Patterns of Coverage

Table 5.4.3 presents the indices of each system's coverage of the SIMS content domain of Algebra. Figure 5.4.4 presents a system-by-subtopic display of the coverage of each subtopic within that domain.

As seen in Table 5.4.3, the median between-system OTL index for Algebra is 74 percent while that for Measurement is 82 percent and for Arithmetic 82 percent. The range of OTL indices for Algebra is much greater than the range in Arithmetic or Measurement. Canada (British Columbia), France, Hungary, Japan, Swaziland and Thailand have OTL

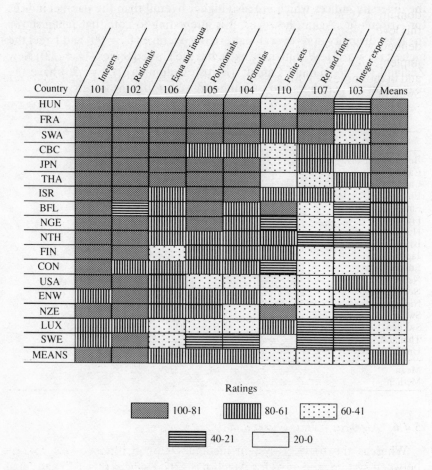

FIG 5.4.4   Population A: Implemented Coverage for Algebra

indices above 80 percent while England and Wales, Finland, Luxembourg, New Zealand and Sweden have OTL indices below 70 percent. As noted in the discussion of Algebra in Chapter 3, these different levels of coverage no doubt reflect the different stages in each school system at which the formal treatment of Algebra is begun as well as the differences in the place within the grade structures of the various systems of the Population A year. The consequences of this pattern of widely different levels of coverage are clearly seen in Figure 5.4.4. Only Integers is widely taught over all of the participating systems although in the block of high coverage (of Algebra) systems, Integers, Linear Equations and Inequalities, Polynomials and Rational Expressions and Formulas and Algebraic Expressions constitute the core of the curriculum. It is noteworthy that two topics included in the SIMS content

The Content of the Implemented Mathematics Curriculum   129

domain, Exponents and Relations and Functions, are not widely taught at
this level. We can conclude that the content domain of Algebra—as this was
defined within SIMS—has a variable and limited isomorphism with the
implemented curriculum of the set of Population A systems. Furthermore,
and in the context of this observation, it is noteworthy that Luxembourg and
Sweden have OTL indices of below 50 percent.

Table 5.4.3 *Population A: Coverage of Algebra*

	Mean	Q3	Md	Q1	Q	U'
Belgium (Flemish)	74	88	80	67	10.5	.10
Canada (British Columbia)	83	93	83	74	9.5	.17
Canada (Ontario)	70	83	73	57	13.0	.15
England and Wales	64	83	67	40	21.5	.19
Finland	70	82	72	58	12.0	.15
France	87	90	83	77	6.5	.18
Hungary	91	93	90	67	13.0	.01
Israel	79	90	80	57	16.5	.22
Japan	83	87	83	80	3.5	.03
Luxembourg	51	73	60	29	22.0	.18
Netherlands	72	82	74	60	11.0	.11
New Zealand	64	77	63	53	17.0	.15
Nigeria	73	NA	NA	NA	NA	NA
Swaziland	87	NA	NA	NA	NA	NA
Sweden	44	60	33	35	12.5	.16
Thailand	82	90	83	77	6.5	.09
United States	67	90	73	53	18.5	.17
Mean	75	84	73	58	13.0	.14
Median	74	87	74	58	12.5	.15

5.4.6 Algebra: Within-system Patterns of Coverage

Whereas the between-system median Q for Arithmetic was 7.0, the
between-system median Q for Algebra is 12.5. France ($Q = 6.7$) and Japan
($Q = 3.4$) again have the lowest indices of variation; Luxembourg ($Q =
22.1$), England and Wales ($Q = 21.7$) and the United States ($Q = 18.4$) have
the highest indices of variation. Figure 5.4.5 presents the I-C Tables from
these low and high variation systems and show very different patterns: in
cases of France and Japan a right-angled I-line drops more or less vertically
to cover half or more of the items in the domain before turning left to
indicate a set of items covered less uniformly; in the high variation cases
the I-line tends to bisect the table and in the cases of England and Wales and
Luxembourg there is a suggestion of plateaus which may indicate tracks
within the system which cover differing amounts of Algebra.

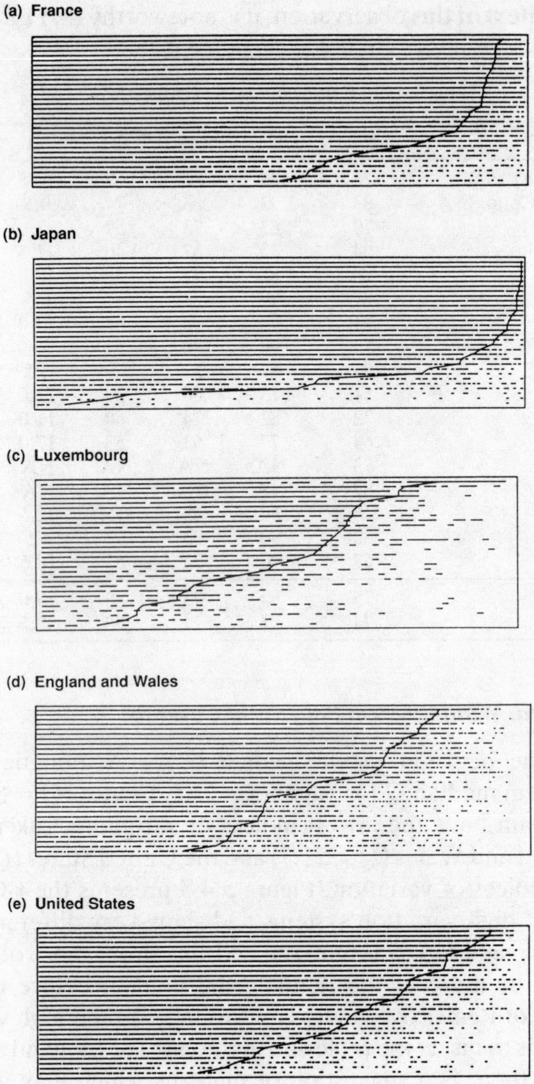

(a) France

(b) Japan

(c) Luxembourg

(d) England and Wales

(e) United States

FIG 5.4.5   Algebra: Illustrative I-C Tables

The diversity indices (U') presented in Table 5.4.3 reinforce this impression of variability in the ways in which the teaching of Algebra is deployed within systems. With the significant exception of Japan, all systems have relatively high indices of diversity and it may be that such diversity is a reflection of the responses that teachers make to the overall variations of readiness for Algebra that they find at this level.

5.4.7　Geometry: Between-system Patterns of Coverage

Table 5.4.4 presents the various indices of implemented coverage of the SIMS content domain of Geometry and Figure 5.4.6 presents a display of coverage of each subtopic within the content domain.

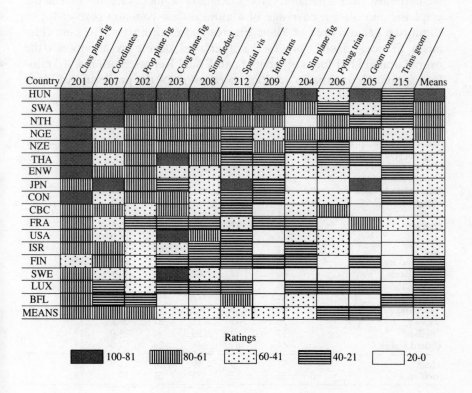

Ratings

100-81　　80-61　　60-41　　40-21　　20-0

FIG 5.4.6　Population A: Implemented Coverage for Geometry

As seen in Table 5.4.4 the between-system median OTL index for Geometry is 51 percent; this contrasts with the median OTL indices for Arithmetic (82 percent) and Measurement (82 percent), and with Algebra where the median coverage index is 74 percent. As was the case in Algebra, but not Arithmetic, the range in levels of coverage is substantial. Hungary has a median OTL index below 35 percent. The causes and consequences of this pattern of substantial variation in coverage are seen in Figure 5.4.6. The topic within Geometry that is most widely taught across systems, Plane Figures, has a between-system coverage index of 62 percent and receives heavy emphasis in only eight of the 16 systems. Furthermore, it would seem as if there are clear differences between clusters of systems in the curricula they offer. England and Wales, and those systems whose curricula are modeled on the English curriculum together with Hungary, Japan, The Netherlands, and Thailand, give Geometry a moderate or substantial emphasis and offer a coverage of a more or less common core of topics within the general area of Plane Figures and Coordinates. Belgium (Flemish), France, Luxembourg, on the other hand, devote less attention to the overall domain but focus on Spatial and Visual Representation and Transformational Geometry, in addition to coverage of the properties of Plane Figures.

Table 5.4.4   *Population A: Coverage of Geometry*

	Mean	Q3	Md	Q1	Q	U'
Belgium (Flemish)	31	33	28	23	5.0	.08
Canada (British Columbia)	50	62	54	41	10.5	.13
Canada (Ontario)	51	67	54	38	17.0	.17
England and Wales	54	67	51	36	15.5	.17
Finland	38	52	34	21	15.5	.21
France	44	49	41	33	8.0	.15
Hungary	87	87	85	56	15.5	.04
Israel	43	51	31	21	15.0	.19
Japan	51	62	51	41	10.5	.10
Luxembourg	35	44	35	23	10.5	.15
Netherlands	66	74	67	60	7.0	.09
New Zealand	60	69	62	49	10.0	.10
Nigeria	64	NA	NA	NA	NA	NA
Swaziland	80	NA	NA	NA	NA	NA
Sweden	36	44	35	27	8.5	.12
Thailand	57	69	56	41	14.0	.11
United States	44	62	46	28	17.0	.14
Mean	52	59	49	34	11.9	.13
Median	51	62	51	33	10.5	.13

5.4.8   Geometry: Within-system Patterns of Coverage

The median between-system index of variability for Geometry is 10.5. Read in the light of the relative low median coverage index for the area this index suggests substantial within-system variation in coverage of Geometry when compared even with Algebra. This may reflect a differential coverage of Geometry in different tracks or courses. The I-C Tables suggest that this may be the case: with the exceptions of systems which have patterns like that seen for Belgium (Flemish) (see Figure 5.4.7(a)), i.e., Belgium (Flemish), France and Luxembourg, the modal table takes the form seen in Figure 5.4.7(b): the I-line bisects the plot but has a suggestion of plateaus that may indicate track effects. It is noteworthy that in all systems a significant number of classes do *not* cover much of the domain of Geometry as defined within SIMS.

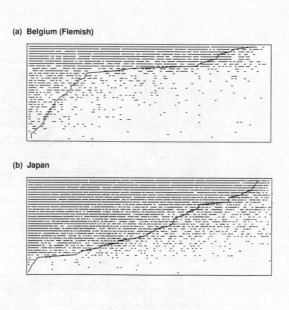

(a) Belgium (Flemish)

(b) Japan

FIG 5.4.7   Geometry: Illustrative I-C Tables

As might be expected, the van der Flier indices U' suggests a pattern of both greater and less diversity in Geometry as compared to Algebra and Arithmetic. In the cluster of moderate and high coverage systems, U' is lower in Canada (British Columbia), England and Wales, and New Zealand and higher in Canada (Ontario), Japan, Thailand and the United States than in these other areas. In general, it would seem that where the median U' is lower than it is in, say, Algebra, the treatment of the area is controlled by more or less fixed assumptions about the scope of coverage, whereas where U' is higher, assumptions about coverage are less firm—and again it may be that the sway of such assumptions are associated with tracking effects.

5.4.9   *Statistics: Between-system Patterns of Coverage*

Table 5.4.5 presents the various indices of implemented coverage for the SIMS content domain of Statistics and Figure 5.4.8 presents a display of coverage within each subtopic in the content domain. It should be noted that Statistics was much less well represented in the SIMS item pool (18 items) than were the other domains.

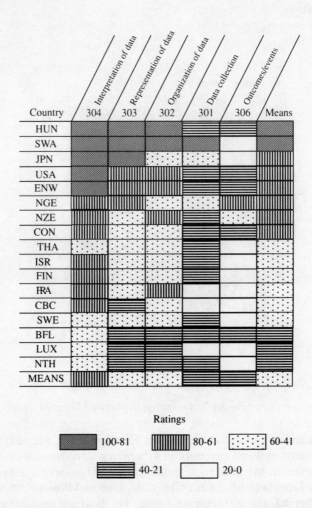

FIG 5.4.8   Population A: Implemented Coverage for Statistics

As seen in Table 5.4.5, the between-system OTL indices for Statistics is 58 percent. This is significantly below the OTL indices for Arithmetic, Measurement and Algebra. Also, as in the case of Geometry, the range in levels of coverage is substantial: Hungary, Japan and Swaziland have OTL indices over 75 percent Belgium (Flemish), Luxembourg and The Netherlands have indices below 40 percent. The consequences of the mixed pattern of coverage is seen in Figure 5.4.9. Only one topic, Interpretation of Data, is widely taught across all systems, although the overall content area as defined by SIMS does seem to match the curriculum of England and Wales, Canada (Ontario), Hungary, and Japan fairly closely.

Table 5.4.5   *Population A: Coverage of Statistics*

	Mean	Q3	Md	Q1	Q	U'
Belgium (Flemish)	38	53	33	17	18.0	.14
Canada (British Columbia)	48	65	47	24	20.5	.15
Canada (Ontario)	60	83	61	39	22.0	.16
England and Wales	69	83	67	44	19.5	.18
Finland	51	69	50	26	21.5	.14
France	51	67	44	28	19.5	.18
Hungary	87	89	83	56	16.5	.03
Israel	52	61	44	17	22.0	.16
Japan	75	89	72	67	11.0	.05
Luxembourg	37	44	28	17	13.5	.13
Netherlands	32	47	21	11	18.0	.13
New Zealand	60	83	67	39	22.0	.18
Nigeria	64	NA	NA	NA	NA	NA
Swaziland	83	NA	NA	NA	NA	NA
Sweden	47	66	43	20	23.0	.17
Thailand	53	83	50	28	27.5	.17
United States	70	89	78	56	16.5	.12
Mean	58	71	53	31	19.4	.14
Median	58	69	50	26	19.5	.15

5.4.10   Statistics: Within-system Patterns of Coverage

The median between-systems index of variability, Q, for Statistics is 19.5. This is by far the largest index of variability across the Population A content domains and when read in the light of the relatively low overall coverage of the domain, implies substantial within-system variation in coverage. In the set of systems with high levels of coverage, i.e., OTL indices over 60 percent, Japan would seem to have the least variation in coverage ($Q = 11.1$); most of the other high coverage systems have Qs above 20. Figure 5.4.9 illustrates a selection of the patterns which underlie the summary statistics seen in Table 5.4.5 and the most noteworthy feature of these tables in the plateaus which occur in many systems' coverage of the domain and variability of coverage seen in some systems, even within the widely taught parts of the subject area.

(a) France

(b) New Zealand

(c) Japan

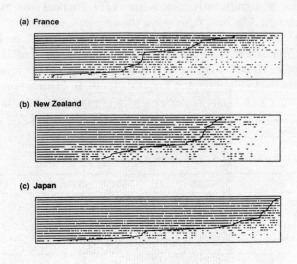

FIG 5.4.9   Statistics: Illustrative I-C Tables

It is interesting to note that, within this pattern, there seems to be little overall variation in the diversity indices; the diversity indices of the higher coverage systems, i.e., OTL indices over 60 percent, fall into a fairly narrow range with the exception, again, of Japan where U' is .05.

5.5   Population A: Between-system Variation in Coverage

Figure 5.5.1 summarizes the coverage data for the Population A content domains across systems. Arithmetic and Measurement tie for first place in the overall comprehensiveness and consistency of coverage across systems but in each of these domains several systems have OTL indices below 80 percent and two systems have OTL indices below 70 percent in each domain. Algebra ranks next in overall coverage but it is interesting to note that in this domain five systems have OTL indices below 70 percent and that Sweden has an OTL index in Algebra below 50 percent. The coverage patterns in Statistics and Geometry, the least comprehensively covered areas, is widely dispersed and in each case a significant group of systems have OTL indices below 50 percent.

Table 5.5.1 summarizes these findings in a different way. The table is ordered by columns (content domains) and rows (systems) and parallels the summary of findings on the intended curricula of the various systems reported in Chapter 4. It is clear that the core curriculum of the Population

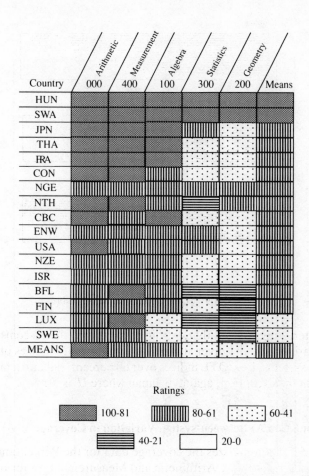

Country	Arithmetic 000	Measurement 400	Algebra 100	Statistics 300	Geometry 200	Means
HUN						
SWA						
JPN						
THA						
FRA						
CON						
NGE						
NTH						
CBC						
ENW						
USA						
NZE						
ISR						
BFL						
FIN						
LUX						
SWE						
MEANS						

Ratings

100-81   80-61   60-41

40-21   20-0

FIG 5.5.1   Population A: Implemented Coverage Across Subject Domains

A grade, as this is reflected in the OTL indices of the various systems, consists of topics from Arithmetic and Measurement. In both Arithmetic and Measurement nine of the 17 systems have OTL indices above 80 percent—but it is interesting to note that these groups vary in their composition. Only six of the 17 systems have OTL indices in Algebra above 80 percent. These findings would seem to have significant implications for the ways in which the achievement profiles from the different systems might be interpreted.

Table 5.5.1   Population A: Indices of Implemented Curricular Coverage

	Measurement	Arithmetic	Algebra	Geometry	Statistics
9	FRA SWA JPN HUN	HUN	HUN		SWA HUN
8	LUX BFL CON NTH THA	NTH USA JPN SWA THA CBC FRA CON	THA CBC JPN FRA SWA	SWA HUN	
7	FIN NZE NGE USA CBC ENW	ISR FIN BFL ENW LUX NGE	CON FIN NTH NGE BFL ISR		USA JPN
6	ISR SWE	SWE NZE	ENW NZE USA	NZE NGE NTH	CON NZE NGE ENW
5			LUX	CBC CON JPN ENW THA	FIN FRA ISR THA
4			SWE	ISR FRA USA	SWE CBC
3				BFL LUX SWE FIN	NTH LUX BFL
2					
1					
0					
N. Items	24	46	30	39	18
Mean	.73	.73	.67	.45	.51

5.6 Population A: Patterns of Within-system Variation

The implications of the findings on between-system differences in the level of coverage for both an analysis of curricula and for the understanding of between-system patterns of achievement are clear. Findings on between-systems variation in the patterns of coverage *within* systems are just as significant: they permit us to address the ways in which a curriculum is *distributed* across schools and classes within a system, a significant issue for its own sake, and they provide a resource for consideration of the *achievement profile* of a system.

The first issue to be considered as we raise these questions is the extent to which within-system variation in coverage varies across systems and content domains within systems. The implication of the earlier discussion of within-system variation is that there are substantial between-system differences in coverage of the different content domains as this is reflected in the semi-interquartile range (Q) of each system's OTL indices. Figure 5.6.1 summarizes these findings using a variant of the tabular format used both in Chapter 4 and this chapter. The entries represent ranges of Qs. The figure is ordered

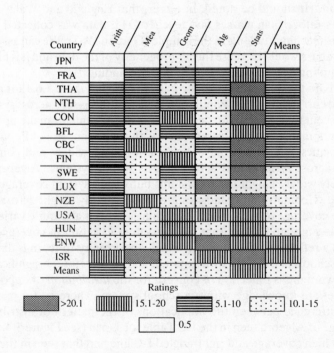

Fɪɢ 5.6.1   Population A: Within-system Variation in Coverage Across
Subject Domains and Systems

by system and content domain from low variation (light shading) to high variation (dark shading).

An inspection of Figure 5.6.1 shows that there are clear differences between both content domains and systems in the extent of variation in within-system coverage. Arithmetic, the most widely-taught domain, has the least within-system variation across most systems while Statistics, with Geometry, the least widely-taught domain, has the greatest variation. Measurement also ranks low in overall variation but it is clear from the figure that the pattern in this domain is very different from that of Arithmetic. (It may be that this is an outcome of the lack of familiarity on the part of Population A teachers with the coverage of lower, often elementary grades.) Algebra ranks fourth in overall variation and, like Measurement, its pattern of variation differs from that seen in Arithmetic, Geometry and Statistics. We will return to the implications of this finding later.

Within the overall pattern of variation across domains, it is also clear that there is a tendency for systems to be somewhat consistent in the overall variability of their within-system patterns of coverage. Japan, France, The Netherlands and Belgium (Flemish) tend to show the least overall variation in coverage while England and Wales and Israel show the greatest overall variation. (It should be noted, however, that England and Wales sampled students rather than classes and teacher OTL data was collected from all teachers teaching students in the sample (Garden 1987). We can assume that this database would increase the heterogeneity of the data and it is likely that this sampling difference is reflected in these findings.)

What are the implications of the pattern seen in Figure 5.6.1 for a view of the core curriculum of the Population A year as this is seen across systems? It would seem that Arithmetic, and perhaps Measurement, make up the only content domains that can be regarded as the common core of Population A mathematics. These are the only areas that are widely and consistently taught across systems and across classes within systems. Algebra is also relatively widely taught across systems but within-system coverage is quite variable. Geometry and Statistics are inconsistently taught across systems and the coverage of these domains within systems is also quite varied.

It was suggested above that within-system variation in coverage can be seen as a reflection of the ways in which opportunity-to-learn is distributed across schools and classes with a system and is, therefore, a significant issue in its own right as reflection of concern for the *distribution* of opportunities for access to knowledge within a school system. There is, for example, a clear difference between the implications of the pattern of distribution of coverage of Algebra seen in the I-C Table for Japan (see Figure 5.4.5 above) with its high coverage and right-angled I-C line and that seen in the parallel I-C Tables for Luxembourg or the United States. In the case of Japan there is substantial uniformity (which might be interpreted as equity) in the instruction being offered within the system. In the cases of Luxembourg and

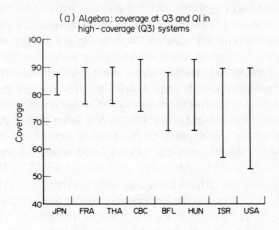

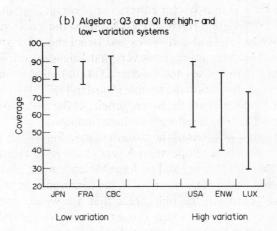

FIG 5.6.2   Algebra: Within-system Variation

the United States there is heterogeneity in the instruction being offered in different classes within each system.

The consequences of these patterns of within-system distribution of instruction in Algebra are highlighted in Figure 5.6.2 which compares the coverage indices for classes at the third and first quartiles of two sets of systems. Figure 5.6.2(a) includes those Population A systems in which the coverage index for high-coverage classes (Q3) is greater than 85 percent and it is clear from the figure that it is the level of coverage offered to low-coverage classes which is the major source of within-system variation in these systems. Figure 5.6.2(b) presents a parallel representation of the distribution of instruction in the three systems with the least within-system variation and the three systems with the greatest within-system variation in Algebra and again it is clear that it is the character of instruction of low-coverage classes which plays a major role in creating within-system variation.

Figure 5.6.3 presents a parallel depiction of within-system variation in Arithmetic, the content domain in which there is the least overall within-system variation. It is clear in these figures that, when compared to Algebra, there is greater uniformity (or equity) in the deployment of instruction across classes within most systems, but even here there is significant variation between systems in the patterns of instruction felt appropriate to

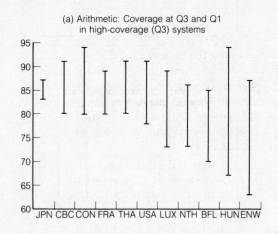

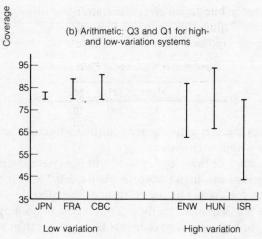

Fig 5.6.3   Arithmetic: Within-system Variation

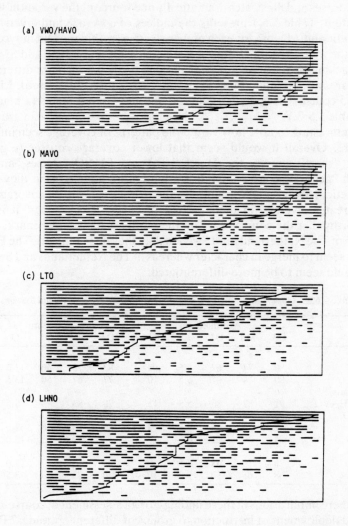

(a) VWO/HAVO

(b) MAVO

(c) LTO

(d) LHNO

FIG 5.6.4   Between-course Variation in Implemented Coverage for Algebra:
The Netherlands

low-coverage classes. In some systems, and Japan is the notable case, assumptions about an appropriate curriculum are more or less constant across all classes within the system whereas in the higher-variation systems there is less such uniformity. And, as was noted above, within-system patterns of variation tend to hold across content domains.

In some systems the obvious source of the kinds of variation seen in the findings we have been considering derives from overt course differentiation at this level. Figure 5.6.4 presents the I-C Tables for Algebra for the four Dutch courses and illustrates dramatically one source of the variation seen in that system. Table 5.6.1 presents the indices of coverage and diversity for Arithmetic and Algebra for each of these courses within the Dutch system as well as the indices for the differentiated courses from Finland and Sweden. Such tracking or streaming of students into courses which are different in character is also found in Finland (where two courses are offered), Luxembourg (5 courses), Sweden (2 courses) and the United States (4 courses). (See Table 3.5.2.)

It is interesting to note how varied the patterns of coverage seen in Table 5.6.1 are. Overall it would seem that lower coverage courses (e.g., the Finnish "short" course, the Dutch LTO and LHNO courses, and the Swedish "general" course) give less attention to Algebra than they do to Arithmetic, have greater variation in the overall patterns of coverage and show greater diversity in their pattern of coverage. However, it is also noteworthy that the courses vary on these indices much more markedly in the case of The Netherlands and Sweden than they do in Finland. The Finnish courses seem to merge in character whereas in The Netherlands and Sweden they would seem to be more differentiated.

Table 5.6.1   *Opportunity-to-Learn in Different Courses Offered Within Systems*

	Arithmetic					Algebra				
	Q3	Md	Q1	Q	U'	Q3	Md	Q1	Q	U'
Finland										
Long	80	74	67	6.5	.09	83	73	60	11.5	.07
Short	80	74	67	6.5	.10	77	67	50	13.5	.14
Netherlands										
VWO/HAVO	91	87	80	5.5	.13	70	83	73	9.5	.07
MAVO	89	83	76	6.5	.14	87	80	67	10.0	.12
LTO	82	74	63	9.5	.13	60	53	38	11.0	.15
LHNO	74	70	61	6.5	.16	57	47	33	12.0	.17
Sweden										
Advanced	78	70	63	7.5	.09	67	57	43	12.0	.16
General	68	61	50	9.0	.10	40	25	13	13.5	.15

Are there implications in these findings for an assessment of course-differentiated deployment of instruction to groups of different aptitudes? Before considering this, another question arises: Does such explicitly-differentiated instruction produce greater variation in coverage than that which is associated with a formal single Population A course? Nine of the 18 systems being considered in this chapter offer only one course at the Population A level; four offer two or more courses differentiated by aptitude and ability; the remaining systems have differentiated courses but the differentiation centers on content (see Chapter 3). Table 5.6.2 rank-orders those systems offering one and two or more aptitude-differentiated courses in the content domains of Arithmetic and Algebra. The systems with differentiated courses and the number of such courses are italicized.

Table 5.6.2   *Population A: Course Organization and Curricular Variation*

Arithmetic	Q	Algebra	Q
Japan	2.2	Japan	3.4
France	4.4	France	6.7
Canada (British Columbia)	5.7	Thailand	7.5
Thailand	6.0	Canada (British Columbia)	9.6
The Netherlands (4)	6.3	*The Netherlands* (4)	9.9
Finland (2)	6.5	New Zealand	11.7
United States (4)	6.5	*Finland* (2)	12.0
Canada (Ontario)	6.6	Canada (Ontario)	13.3
Luxembourg (5)	7.9	*Sweden* (2)	13.4
New Zealand	8.7	Israel	16.7
Sweden (2)	9.0	*United States* (4)	18.4
Israel	18.5	*Luxembourg* (4)	22.1
Median Q for all systems	7.0	Median Q for all systems	12.6

Inspection of Table 5.6.2 makes it clear that explicit differentiation of Population A courses is not unequivocally associated with greater differentiation of coverage than that found in systems which offer only one course at this level. This absence of an effect from differentiation is seen in Arithmetic, the content domain which is a major part of the core of the Population A curriculum. However, in Algebra there is some tendency for differentiated systems to show higher variation than do systems with only one course. But it is also clear that in both domains factors other than formal course differentiation as such are at work. Thus, within systems with a common course there are different structures of course development and/or prescriptions possible, e.g., common core plus elective/enrichment topics as well as the *de facto* differentiation often associated with tracking and setting. But, despite this, Table 5.6.2 would seem to raise an intriguing cluster of questions associated with the issue of course differentiation. Why do some systems formally differentiate at the Population A level while other systems seem to see no need for such differentiation? If, as may be the case, such differentiation is associated with greater variation in between-class coverage, why is this seen as desirable in some systems and not in others? And, finally, what are the consequences of such differentiation for the distribution of opportunity-to-learn?—for achievement? and for the larger issue of educational opportunity within the school?

The effects of course differentiation on achievement are matters for later volumes in this series. We can only consider here the issue of the sources of such differentiation, particularly in Algebra. And in this regard, it is noteworthy that those systems which have higher variation in their teaching of Algebra also tend to be systems with lower levels of coverage of Algebra (see Table 5.5.1 above). Variation and differentiation are, therefore, perhaps manifestations of underlying assumptions about the scope and sequence of the overall curriculum at the Population A grade level; it would seem that in such high variation-low coverage systems as Luxembourg, Sweden or the United States, Algebra is simply less central to the core curriculum of the Population A year than it is to the curricula of the low variation-high coverage systems.

Why might this be the case? In Chapter 3 above, it was suggested that the Population A year of the various systems sampled different points in the structure of the various school systems. It is perhaps significant that high variation-low coverage systems seen in Table 5.6.2, i.e., Canada (Ontario), Israel, Luxembourg, Sweden and the United States, have Population A years which are either located within the elementary school system or in the first year of the secondary system. Furthermore, three of these systems (Canada (Ontario), Israel and the United States) emphasize local control in

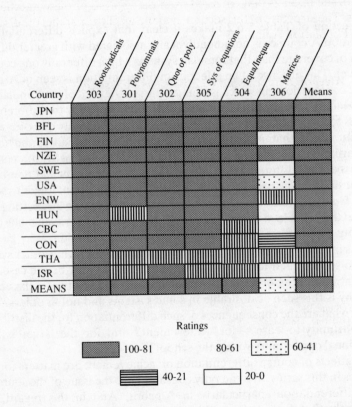

Fig 5.7.1 Population B: Implemented Coverage for Algebra

their administration of the curriculum and Luxembourg, Sweden and the United States have offered multiple courses at this level. While *some* classes in these systems are offered courses with a substantial emphasis on Algebra, many classes are not offered such courses—because, perhaps, Algebra is not seen as a canonical part of the curriculum of an upper elementary-first year secondary Population A year. It is, for many teachers and schools, discretionary content which has a place in the experiences of some but not all Population A students.

5.7 Population B: Patterns of Content Coverage

Twelve systems administered the SIMS *Teacher OTL Questionnaire* as part of their Population B testing. In the pages that follow we will consider some of the findings that emerged from the analysis of these questionnaire results. The structure of this section, and the displays, parallel the discussion of the Population A findings.

5.7.1 Algebra: Between- and Within-system Patterns of Coverage

Table 5.7.1 presents the OTL indices for the content area of Algebra and Figure 5.7.1 presents the coverages of topical areas within the overall domain. It is clear that there is a substantial consensus about the scope of coverage of Algebra across all systems except Israel. There is likewise little within-system variation in levels of coverage in all systems except Israel and Thailand. There would also appear to be little diversity in within-systems patterns of coverage. It is noteworthy, however, that England and Wales, Israel, Canada (Ontario), Thailand and the United States have much higher diversity indices than the other systems.

Table 5.7.1   *Population B: Coverage of Algebra*

	Mean	Q3	Md	Q1	Q	U'
Belgium (Flemish)	92	100	92	88	6.0	.07
Canada (British Columbia)	83	88	84	76	6.0	.06
Canada (Ontario)	83	85	80	72	6.5	.08
England and Wales	87	96	88	80	8.0	.12
Finland	92	96	92	88	4.0	0
Hungary	86	92	84	83	4.5	.04
Israel	72	76	60	26	25.0	.16
Japan	100	100	100	100	0	0
New Zealand	92	100	96	88	6.0	.08
Sweden	90	96	91	84	6.0	.04
Thailand	78	88	80	64	12.0	.18
United States	89	92	88	84	4.0	.10
Mean	87	92	86	77	7.3	.08
Median	87	96	88	84	6.0	.08

5.7.2   *Elementary Functions and Calculus: Between-system Patterns of Coverage*

Table 5.7.2 presents the OTL indices for the content domain of Elementary Functions and Calculus and Figure 5.7.2 presents the coverage of topical areas within this subject domain. The most noteworthy feature of the table is, of course, the difference between the coverage indices for Canada (British Columbia) (Md = 30 percent) and the United States (Md = 50 percent) and most of the other systems (which have coverage indices above 85 percent). Hungary (Md = 67 percent), Israel (Md = 76 percent) and Thailand (Md = 63 percent) fall between these poles.

It is clear from an inspection of Figure 5.7.2 that there is a more or less common set of topics being covered within the high and moderate coverage

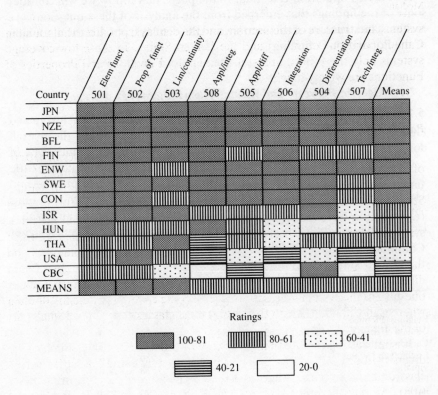

FIG 5.7.2   Population B: Implemented Coverage for Elementary Functions and Calculus

Table 5.7.2   *Population B: Coverage of Elementary Functions and Calculus*

	Mean	Q3	Md	Q1	Q	U'
Belgium (Flemish)	89	94	89	85	4.5	.09
Canada (British Columbia)	35	39	30	26	6.5	.04
Canada (Ontario)	83	87	85	76	5.5	.13
England and Wales	88	96	91	83	6.5	.09
Finland	88	96	94	89	3.3	.02
Hungary	68	77	67	55	11.0	.05
Israel	79	87	76	59	14.0	.16
Japan	94	100	98	94	3.0	.01
New Zealand	94	98	94	91	3.5	.07
Sweden	88	91	87	81	5.0	.06
Thailand	66	70	63	52	9.0	.09
United States	58	85	50	33	26.0	.08
Mean	78	85	77	69	8.2	.07
Median	88	91	87	81	6.5	.08

systems and that this curriculum is very different in scope from that found in Canada (British Columbia) and the United States. In these low-coverage systems only the topical areas of Elementary Functions and Properties of Functions are widely taught.

5.7.3   Elementary Functions and Calculus: Within-system Patterns of Coverage

An inspection of Table 5.7.2 suggests that in addition to an overall pattern of high coverage, there is little variation in coverage of Elementary Functions and Calculus in most systems. The significant exception is the United States with a variability index (Q) of 26.4, but Hungary ($Q = 11.0$) and Israel (Q 14.2) also have much higher indices of variation than most of the other systems. Figure 5.7.3 presents the I-C tables from Finland, Canada (British Columbia) and the United States. In both Canada (British Columbia) and the United States, in contrast to Finland, the I-lines bisect the tables but in different ways; in the case of the United States it would seem that about one-fifth of the system's classes do receive some coverage of the full domain whereas in Canada (British Columbia) most classes cover only a small part of the area.

It is interesting to note that in the case of Canada (British Columbia) the diversity index U' is low (.04), suggesting substantial similarity across classrooms in coverage, whereas in the United States U' is somewhat higher (.08). Among the other systems, Israel ($U' = .16$) and Canada (Ontario) ($U' = .13$) have far higher diversity indices than any other system.

5.7.4   Number Systems: Between-system Patterns of Coverage

Table 5.7.3 presents the indices of coverage for the content domain of Number Systems and Figure 5.7.4 presents the coverage of topical areas

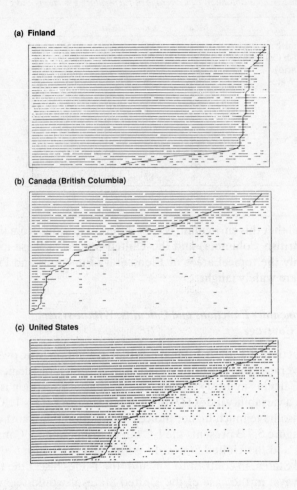

(a) Finland

(b) Canada (British Columbia)

(c) United States

FIG 5.7.3   Elementary Functions and Calculus: Illustrative I-C Tables

within the domain. Belgium (Flemish), England and Wales, Finland, Japan, New Zealand, Sweden and the United States have median OTL indices over 80 percent, suggesting substantial overall coverage of this domain as it was defined within the SIMS item pool. However, Hungary, Israel and Canada (Ontario) have OTL indices below 65 percent. Inspection of Figure 5.7.4 suggests that there is a high overall coverage of all subtopics in most systems but it is interesting to note that Japan among the high coverage systems and Israel and Canada (Ontario) among the lower coverage systems appear to give less emphasis than do the other systems to Complex Numbers.

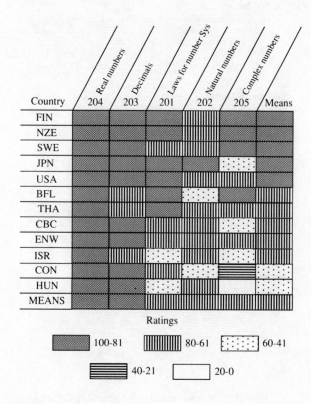

FIG 5.7.4   Population B: Implemented Coverage for Number Systems

5.7.5   *Number Systems: Within-system Patterns of Coverage*

When compared to both Algebra and Elementary Functions and Calculus, Number Systems has a much higher overall within-system variation in coverage. The median Q for Algebra is 6.0 and for Elementary Functions and Calculus 5.9 whereas Q for Number Systems is 11.2. There is, however, substantial between-system variation in these indices. Canada (British Columbia) (8.0), Finland (8.0), Japan (2.6), New Zealand (8.0), and Sweden (6.5) have relatively low indices of variation whereas England and Wales (16.0), Israel (18.0), and Thailand (18.4) have much higher Qs.

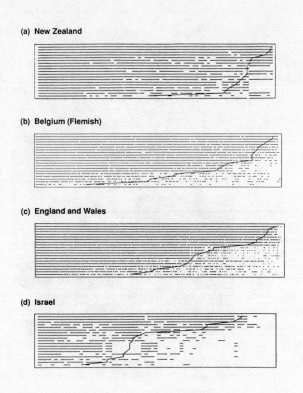

(a) New Zealand

(b) Belgium (Flemish)

(c) England and Wales

(d) Israel

FIG 5.7.5   Number Systems: Illustrative I-C Tables

Figure 5.7.5 presents some typical I-C Tables for this area. In the case of New Zealand we see a table representing almost universal coverage of the content domain whereas the tables for Belgium (Flemish) ($Q = 11.0$) and England and Wales ($Q = 16.0$) would seem to suggest some tracking of coverage across different classes in each sample. In the case of Israel ($Q = 18.0$), where there is relatively low coverage of Number Systems, such tracking would seem to be clear.

The diversity indices for Number Systems show a similar mixed pattern to that seen in the indices of variation. Hungary ($U' = .02$), Japan ($U' = 0$) and Sweden ($U' = .04$) have very low indices suggesting little diversity from the modal pattern of variation found in each case. These are, moreover, systems mostly with relatively high coverage of the content domain. England and Wales ($U' = .12$), Israel ($U' = .18$), New Zealand ($U' = .14$), Thailand ($U' = .21$) and the United States ($U' = .11$) have much higher indices of diversity.

Table 5.7.3   *Population B: Coverage of Number Systems*

	Mean	Q3	Md	Q1	Q	U'
Belgium (Flemish)	78	90	79	68	11.0	.10
Canada (British Columbia)	74	84	74	68	8.0	.07
Canada (Ontario)	60	76	64	53	11.5	.08
England and Wales	74	90	79	58	16.0	.12
Finland	88	95	90	79	8.0	.10
Hungary	56	67	54	48	9.5	.02
Israel	64	68	53	32	18.0	.18
Japan	82	84	79	79	2.5	00
New Zealand	88	95	90	80	9.0	.14
Sweden	87	90	84	77	6.5	.04
Thailand	75	90	74	53	18.5	.21
United States	81	90	84	74	8.0	.11
Mean	76	85	75	64	10.5	.10
Median	78	90	79	68	9.5	.10

5.7.6   Geometry: Between-system Patterns of Coverage

Table 5.7.4 presents the coverage indices for the content domain of Geometry and Figure 5.7.6 presents the coverages of subtopics within the domain. Overall the coverage of Geometry is only moderate (Md = 64 percent) when compared to the other Population B domains. There is, moreover, a substantial range in coverage indices. Japan has a median coverage index of 86 percent while Canada (British Columbia), Canada (Ontario) and Israel have median coverage indices below 45 percent.

As seen in Figure 5.7.6, this range in levels of coverage reflects widely different patterns of coverage of the subtopics included within the SIMS item pool in Geometry. The nine items in Trigonometry have a mean OTL index of 86 percent and were reported as widely taught in all systems, but outside this core, coverage indices drop markedly. The overall coverage index for the six items in Analytical Geometry is 69 percent with Canada

Table 5.7.4   *Population B: Coverage of Geometry*

	Mean	Q3	Md	Q1	Q	U'
Belgium (Flemish)	79	89	79	68	10.5	.08
Canada (British Columbia)	44	54	43	39	7.5	.03
Canada (Ontario)	52	46	43	36	5.0	.10
England and Wales	64	71	64	54	8.5	.08
Finland	72	75	71	68	3.5	.03
Hungary	65	72	65	55	7.5	.03
Israel	43	44	36	25	9.5	.03
Japan	85	89	86	82	3.5	.01
New Zealand	68	75	68	60	7.5	.12
Sweden	61	69	58	51	6.5	.03
Thailand	62	68	57	46	11.0	.07
United States	54	61	54	43	9.0	.05
Mean	62	68	60	52	7.5	.06
Median	64	71	64	54	7.5	.05

(British Columbia) and Israel having coverage indices of about 50 percent while Japan has a coverage index of 96 percent. For the seven items in Vector Methods the mean coverage index is 54 percent and the United States, Canada (British Columbia), Canada (Ontario) and Israel report a coverage of less than 40 percent of the items. The five items in Transformational Geometry have a mean coverage index of 24 percent and reflect content which is virtually unrepresented in the curricula of most systems. It would seem that, as was the case with Population A Geometry, the SIMS items in Geometry do not map up at all well on to the curricula of most of the participating systems although there is a core within the area in which there is a substantial international consensus.

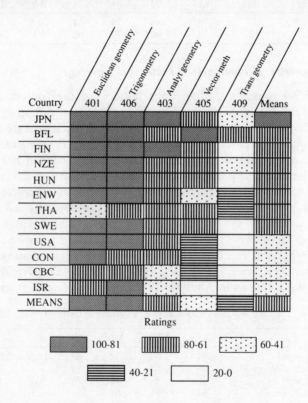

Fig 5.7.6   Population B: Implemented Coverage for Geometry

5.7.7   *Geometry: Within-system Patterns of Coverage*

As was the case with the areas of Algebra and Elementary Functions and Calculus, the median index of variation (Q) for the content domain of Geometry was comparatively low, 7.9. In the cases of Finland and Japan, which are among the systems with the highest levels of coverage, Q is under 3.5. In the set of systems with moderate levels of coverage it is interesting to note that the indices of variation are comparatively high; and as seen in Figure 5.7.7, this relatively high degree of variation seems to be associated with plateaus in the I-lines suggesting some tracking of content to different classes within the system. In the case of England and Wales, this plateauing is associated with sharp divisions between three kinds of classes. In the case of New Zealand and the United States the plateaus seem to be associated with types of classes but the patterns seem less coherent.

It is interesting to note that New Zealand also has the highest diversity index ($U' = .12$), followed by Belgium (Flemish) ($U' = .08$) and England and Wales ($U' = .08$). In all other systems the diversity indices are low.

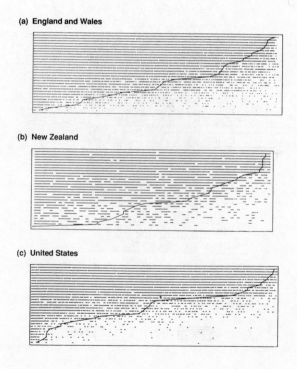

(a) **England and Wales**

(b) **New Zealand**

(c) **United States**

FIG 5.7.7   Geometry: Illustrative I-C Tables

5.7.8   *Sets and Relations, Probability and Statistics, and Finite Mathematics*

The remaining content areas within the SIMS Population B grid are represented by comparatively few items; Sets and Relations, and Probability and Statistics, by seven items each and Finite Mathematics by four items. The emphasis implied by this weighting reflected, of course, the judgments made in the earliest phases of the study about the relative importance of these areas in the international Population B curriculum. And in the cases of Probability and Statistics and Finite Mathematics even this level of emphasis

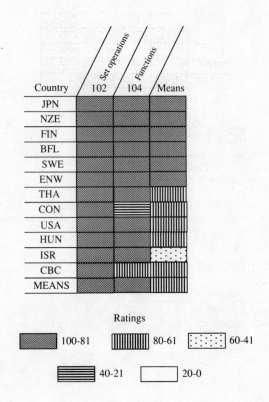

Fɪɢ 5.7.8   Population B: Implemented Coverage for Sets and Relations

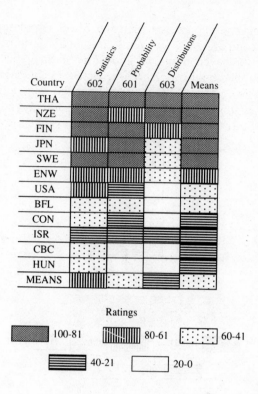

FIG 5.7.9   Population B: Implemented Coverage for Probability and
Statistics

Table 5.7.5   *Population B: Coverage of Sets and Relations*

	Mean	Q3	Md	Q1	Q
Belgium (Flemish)	91	100	86	86	7.0
Canada (British Columbia)	66	86	71	57	14.5
Canada (Ontario)	62	71	62	43	14.0
England and Wales	54	85	57	27	28.0
Finland	88	100	86	71	14.5
Hungary	43	63	43	29	17.0
Israel	38	43	29	7	18.0
Japan	95	100	100	86	7.0
New Zealand	85	100	86	71	14.5
Sweden	63	71	57	43	14.0
Thailand	79	100	86	71	14.5
United States	83	100	86	71	14.5
Mean	71	85	71	55	14.8
Median	79	100	86	71	14.5

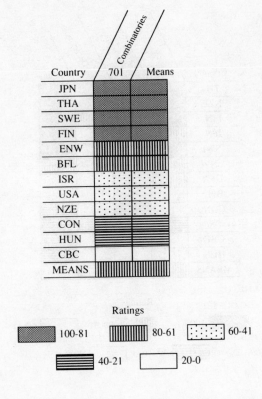

FIG 5.7.10   Population B: Implemented Coverage for Finite Mathematics

was less a reflection of the role that these areas were seen to have in current curricula and more a reflection of the interest that they might have as new topics within future curricula. Tables 5.7.5 and 5.7.6 present the indices of coverage for these areas and Figures 5.7.8 to 5.7.10 present the topical emphases within them.

The median OTL index for Sets and Relations is 79 percent, but within the high overall emphasis on the area, there are clear between-system differences. Hungary (43 percent) and Israel (29 percent) give the content domain much less attention than, say, Japan, Belgium (Flemish), Finland, New Zealand, Thailand, and the United States where the median OTL indices are over 85 percent. These different levels of coverage are reflected in the topical coverages seen within the overall content area. Two items in the domain were drawn from the topic of Set Operations and five from Functions. High coverage systems tended to emphasize both the areas while lower coverage systems gave each less emphasis.

Table 5.7.6 *Population B: Coverage of Probability and Statistics*

	Mean	Q3	Md	Q1	Q
Belgium (Flemish)	44	71	43	14	28.5
Canada (British Columbia)	28	29	29	14	7.5
Canada (Ontario)	33	38	29	17	10.5
England and Wales	71	100	71	43	28.5
Finland	85	100	100	86	7.0
Hungary	26	35	29	15	10.0
Israel	30	14	0	0	7.0
Japan	82	100	100	86	7.0
New Zealand	86	100	100	71	14.5
Sweden	81	100	80	59	20.5
Thailand	91	100	86	71	14.5
United States	46	71	29	29	21.0
Mean	58	72	58	42	14.7
Median	71	100	71	43	14.5

The seven items defining the area of Probability and Statistics (3 each for Probability and Statistics and one item for Distributions) have a median OTL index of 57 percent but there is, again, a wide variation in levels of coverage. Finland, Japan, New Zealand, Sweden and Thailand have coverage indices over 80 percent while Canada (British Columbia), Canada (Ontario), and Hungary have OTL indices below 40 percent; Israel reported no coverage of the domain.

As seen in Figure 5.7.9, part of the difference between systems in levels of coverage of Probability and Statistics is accounted for by their differential treatment of the subtopics that make up the domain. High-coverage systems such as Thailand, Finland, Sweden, and England and Wales cover both areas more or less equally; New Zealand and the United States give more attention to Statistics, while Japan gives more attention to Probability.

The four items defining the content domain of Finite Mathematics (Combinatorics) again shows a wide range in levels of coverage. Belgium

Table 5.7.7 *Population B: Coverage of Finite Mathematics*

	Mean	Q3	Md	Q1	Q
Belgium (Flemish)	63	100	100	0	50.0
Canada (British Columbia)	10	25	0	0	12.5
Canada (Ontario)	40	50	34	19	15.5
England and Wales	65	100	75	25	37.5
Finland	83	100	100	75	12.5
Hungary	29	47	25	25	11.0
Israel	56	100	25	0	50
Japan	99	100	100	100	0
New Zealand	51	75	50	25	25.0
Sweden	89	100	100	75	12.5
Thailand	92	100	100	75	12.5
United States	55	100	50	25	37.5
Mean	61	83	63	37	23.0
Median	63	100	75	25	15.5

(Flemish), Finland, Japan, Sweden and Thailand have OTL indices of 100 percent whereas Canada (British Columbia), Canada (Ontario), Hungary and Israel have indices below 40 percent.

5.8   Population B: Between-system Variation in Coverage

Table 5.8.1 summarizes the coverage data for the Population B domains across systems. Algebra ranks first in the comprehensiveness and consistency of coverage across systems; 10 of the 12 systems have OTL indices over 80 percent in this domain. Elementary Functions and Calculus and Number Systems rank second and third in overall coverage with most systems having OTL indices above 70 percent, but in both of these areas there is a significant subset of systems with coverage indices below 70 percent and a cluster of systems with OTL indices below 60 percent. The coverage pattern in Sets and Relations is widely dispersed across systems: one system has an OTL index over 90 percent, three have OTL indices between 80 percent and 90 percent, three indices between 70 percent and 80 percent, two indices between 60 percent and 70 percent, one an index between 50 percent and 60 percent and one an OTL index below 50 percent. In Geometry the overall level of coverage is significantly lower, and the indices for individual systems range from over 70 percent (three systems) to below 50 percent (two systems). Probability and Statistics shows a bimodal pattern with five systems having OTL indices above 80 percent and six systems having OTL indices below 50 percent.

Figure 5.8.1 summarizes the findings from the OTL Questionnaire in a different way and directs attention to the complex problem of the match between the SIMS item pool and each system's implemented curriculum. Across the domains of Algebra, Elementary Functions and Calculus, Number Systems and Sets and Relations, the core of consistently-covered areas within the pool, only Japan, New Zealand and Finland have high coverage of *all* of these domains. Hungary and Canada (British Columbia) have a consistently low level of coverage of all of these domains except Algebra and Israel has a low coverage of all domains. Other systems show inconsistent patterns of coverage of the different domains. These findings have obvious implications for both a view of curriculum of mathematics at this level and for the ways in which the achievement profiles of the different systems might be interpreted.

5.9   Population B: Patterns of Within-system Variation

It was suggested in the discussion of within-system variation in content coverage at the Population A level (Section 5.6 above) that such variation can be seen as an indicator of the ways in which a curriculum is distributed across the schools and classes within an educational system. Such differences in patterns of distribution have obvious implications for understanding

Table 5.8.1   Population B: Indices of Implemented Curricular Coverage (Teacher Opportunity-to-Learn)

	Algebra 300	Elementary functions and calculus 500	Number systems 200	Sets and relations 100	Geometry 400	Finite mathematics 700	Probability and statistics 600
10	JPN						
9	SWE BFL FIN	JPN NZE					
8	CBC CON HUN FIN NZE	CON ENW FIN SWE BFL	USA JPN SWE FIN NZE	NZE FIN BFL		THA JPN	THA
7	ISR THA	ISR	CBC ENW THA BFL	CON THA USA	FIN BFL	FIN SWE	SWE JPN FIN NZE
6		THA HUN	CON ISR	CBC SWE	SWE THA ENW	BFL ENW	ENW
5		USA	HUN	HUN	CON USA	NZE USA ISR	
4				ISR ENW	ISR CBC	CON	BFL USA
3		CBC				HUN	ISR CON
2						CBC	HUN CBC
1							
0							
N. Items	25	46	19	7	28	4	7
Mean	.87	.78	.76	.71	.62	.61	.59

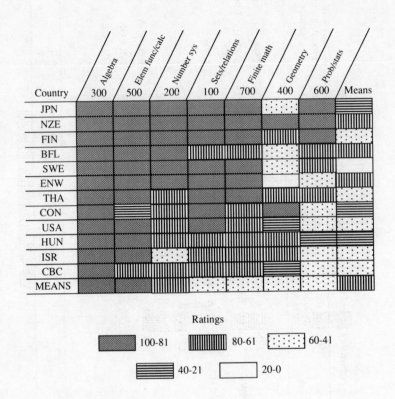

FIG 5.8.1   Population B: Implemented Coverage Across Subject Domains

patterns of achievement as well as posing interesting questions about the curricular assumptions of school systems. In this section we will explore these issues again as they emerge in the Population B component of SIMS.

Figure 5.9.1 offers representation of within-system variation in content coverage across systems and content domains. The figure parallels Figure 5.6.1 above: the shading represents bands of within-system semi-interquartile ranges of teacher OTL and the figure is ordered by both the overall variation found within each system (*Q*) and the between-system median index of variation for each content domain. Light shading represents low variation and dark shading high variation.

As was the case with the Population A year it is immediately clear on inspection of Figure 5.9.1 that there is substantial variation across both

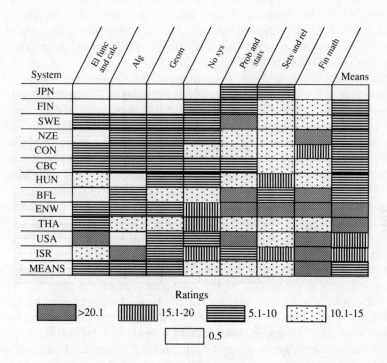

FIG 5.9.1   Population B: Within-system Variation in Coverage Across
Subject Domains and Systems

content domains and systems in the extent of variation in within-system coverage. Elementary Functions and Calculus and Algebra, the most widely taught domains, show the least variation while Finite Mathematics, Sets and Relations, and Probability and Statistics show the greatest variation. These latter areas are, however, the least widely taught overall and represented by comparatively few items in the SIMS pool. Number Systems, the third ranked domain in coverage (sampled by 19 items), also shows a pattern of significant variation.

Within this pattern of variable variation across domains, it is also clear that there is a tendency for systems to be more or less consistent in the overall patterns of within-system coverage—although the overall variability is less marked than it was in the case of Population A. But within this overall pattern, Japan and Finland show the least variation in overall coverage while

Israel, England and Wales, Thailand and the United States show the greatest variation. However, as was noted above, in the case of England and Wales the pattern of variation which emerged from that system may be attributable to the sampling design used in the national study (Garden 1986). In the case of both Israel and the United States the variation seen in Elementary Functions and Calculus (the United States) and Algebra (Israel) is especially noteworthy.

What are the implications of the pattern of variation seen in Figure 5.9.1 for our view of the core curriculum of the Population B year as this is seen cross-culturally? It would seem that Algebra represents the only content domain that is truly common across *all* systems and across all classes within systems—although, as was noted above, Elementary Functions and Calculus is also very widely and comprehensively taught in most Population B systems. Sets and Relations, Probability and Statistics and Finite Mathematics are much less widely taught and are associated with significant variation in coverage within the various systems. Geometry is not comprehensively covered by most systems, but within the coverage of the domain that is attempted by the various systems there is substantial within-system uniformity.

In the case of Population A, our discussion of within-system variation concluded with a discussion of the consequences of the distribution of coverage within a system for the distribution of opportunity to learn. Although we noted above that this issue would also seem to be present in the content domain of Number Systems and Geometry, it did not emerge as a significant issue in Algebra in any system and in Elementary Functions and Calculus only emerged as a major phenomenon in the case of the United States and as a lesser problem in the cases of Canada (Ontario), Hungary, Israel and Thailand. It is perhaps significant that in Canada (Ontario), Hungary and the United States, Population B is made up of students in a number of different courses—with the United States being the extreme case in its availability of five courses at this level and only one of these courses (enrolling 20 percent of the Population B sample of classes) included a substantial component of the calculus. (McKnight et al. 1987). In the cases of the other systems in which there is significant variation seen in Figure 5.9.1 a variety of course-taking patterns occurs in Belgium (Flemish) and England and Wales (see Chapter 3) and variety in course-patterns around a common core of content coverage is also found in Canada (Ontario), Hungary and Sweden. However, while these structural differences do seem to create some variation in within-system coverage in these systems, it is clear that the coverage of the core content of Population B mathematics is more or less consistent across most classrooms within each system—with the exception of the United States. As we saw in the discussion in Chapter 3 of the issue of curriculum control, it is the United States that might be expected to show the most variation at this level because of its structure of multiple courses within a system of decentralized (i.e., local) control of content selection. The findings of this discussion would seem consistent with that expectation. Although we also hypothesized there that a parallel pattern might be found in the case of Canada (Ontario), it was also suggested the consequences of the variety of course offerings and decentralized control might be muted in that system by the recent emergence of such a control structure. This would seem to be the case. In the remaining systems the existence of a more or less firmly prescribed curriculum at the Population B level and more or less firm centralized control of "standards" by way of either external examinations or external monitoring of assessment seems to result in substantial equity in within-system content coverage—albeit one that is modulated in some systems by the variation in coverage in such domains as Number Systems, Geometry, Sets and Relations and Probability and Statistics.

5.10 Summary

This chapter has sought to describe the implemented curriculum of the set of SIMS systems that incorporated the Teacher Opportunity-to-Learn Questionnaire in their component of the Study. This instrument asked the teachers of the classes included in each system's sample to describe their class's coverage of each item in the SIMS item pools. The responses to this instrument provide a basis for a description of each system's implemented curriculum.

What do the findings that emerge from the responses to the teacher OTL questionnaire suggest about the curricula of the participating systems? First it would seem that a derived index of implemented coverage provides a better basis for prediction of students in most systems than a parallel index of the intended curriculum. This finding may be interpreted as offering general support for a claim that teacher responses to the OTL questionnaire offer a generally valid measure of the actual curriculum-in-use in most systems. Second, there is between- and within-system diversity at both the Population A and Population B levels in the curricula-in-use in the various systems. At the Population A level the core of topics that receives near universal coverage is confined to topics in Arithmetic (particularly Common Fractions), Dimensional Analysis and Decimal Fractions, Measurement (particularly Standards Units and Approximation) and within Algebra the topics of Rationals and Integers. At the Population B level the core of topics that receives near universal coverage is large: in Algebra, Roots and Radicals, Polynomials, Quotients of Polynomials, Equations and Inequalities and Systems of Equations and Inequalities; in Elementary Functions and Calculus, virtually all topics with the content domain were intensively covered by the systems with a high coverage of the domain while the low-coverage systems gave moderate attention to only the topics of Elementary

Functions, Properties of Functions, and Limits and Continuity; within Geometry, only Euclidean Geometry and Trigonometry received significant attention in most systems.

The opportunity-to-learn measure also provides a basis for within-system analysis of variation in coverage across the classrooms within a system. At the Population A level, substantial within-system variation emerges as a result of the formal or *de facto* tracking found in many systems. The effect of such tracking is to further reinforce the role of Arithmetic and Measurement as the core areas within the common Population A curriculum. For Population B, much less within-system variation emerges, a reflection, it would seem, of both the strong effect of the examination structures found at this level in most systems and a broad consensus among the teachers of the goals to be achieved by their classes of mathematical specialists.

6

Outputs and Outcomes of Mathematics Education

6.1 Introduction

The report of the Cockcroft Committee on the teaching of mathematics in English schools is entitled *Mathematics Counts* (1982). The pun in this title is clearly intended to attract attention, but the theme of the importance of mathematics education does determine the shape of the report. Mathematics *does* count—and counts more and more as increasing numbers of occupations become more and more mathematically-based and as social and technological development are seen to be linked to technological modernization through the application of science and mathematics to industry and commerce. In the words of *Mathematics Counts*:

. . . Mathematics is fundamental to the study of the physical sciences and of engineering of all kinds. It is increasingly being used in medicine and the biological sciences, in geography and economics, in business and management studies. It is essential to the operations of industry and commerce in both office and workshop. (p. 2)

In recent years a growing recognition of the importance of mathematics has led to a widespread concern in many societies about the *output* of mathematics learning that emerges from educational systems. Mathematics is, or is seen to be, of such importance that the quality and quantity of mathematics teaching and learning offered by school systems has become, to use Davis's (1984) term, a *social issue*—with the common assumption being that in most school systems more students need to be taken further in their study of mathematics. This conclusion lies at the base of the theme of "mathematics for all," that is thought by many to be *the* theme for mathematics education for the balance of this century. But, in the schools, questions about the possibility of a mathematics for all remain unanswered for many. (Damerow et al. 1986; Howson, Nebres and Wilson 1985; Howson and Wilson 1986.) How many students *can* be taken how far? Seen from within the field of mathematics education this question often seems two-pronged: How many students can be taken how far, given the "standards" that are appropriate to the subject? A concern for quantity sits uneasily alongside a concern for quality. A concern for the many sits uneasily alongside a concern for the talented.

167

Issues of these kinds typically arise *within* school systems and are considered within frameworks that are profoundly colored by the assumptions of existing practice. The way things are becomes a datum against which proposals for different futures are measured. It is the task of a comparative study like SIMS to call just such datums into question. Thus, as has been seen in Chapter 3 (Table 3.6.1), there are substantial differences between systems in the proportions of their age cohorts involved in advanced mathematics in the terminal grades of secondary schools. When read alongside the implemented coverage data reported in Chapter 5, these differences raise a set of important questions about what is being "achieved" in different systems.

Such questions define the theme of this chapter. Attention will center on two aspects of the *yield* of mathematics instruction in a subset of the Population B systems—those for which implemented coverage data are available, viz., Belgium (Flemish), Canada (British Columbia), Canada (Ontario), England and Wales, Finland, Hungary, Israel, Japan, New Zealand, Sweden, and the United States. The general concern of this chapter focuses on the differences that are found among the systems in the *yield* of mathematics education. We will also consider a crucial subtheme within this larger topic: the differences among systems in the participation of girls in advanced mathematics.

6.2 Yield

As we have suggested, there are significant differences between the SIMS systems in the proportions of their age cohorts enrolled in Population B courses. But, there are also differences in the coverage (and, as will be seen in Volume 2, in the attainment) of systems. These components of outcomes obviously have potential for interactions and, when combined, may be thought of as constituting the *yield of mathematics teaching* that emerges from an educational system.

A conception of yield was introduced into the IEA studies by Postlethwaite (1967) in his reanalysis of the data derived from FIMS—but he used the notion in a somewhat different way than is found here. The design of that study enabled Postlethwaite to measure three aspects of the *attainment* of students in the participating systems: (1) that associated with students in the terminal grade who were specialists in mathematics and science (FIMS Population 3a); (2) that associated with students in the terminal grade who were *not* specialists in mathematics and science (FIMS Population 3b); and (3) that associated with both groups combined. By combining the proportions of the relevant age cohorts making up these groups, Postlethwaite was able to address the question, "How many are taken how far?" He found that different systems have marked differences in the yields of learning being produced. One formulation of these differences is presented in Figure 6.2.1.

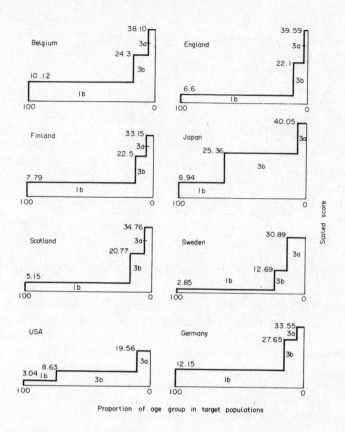

Proportion of age group in target populations

FIG 6.2.1   FIMS: Outcomes of Mathematics Teaching

In this chapter we focus on the issue of yield as it surfaces within the context of mathematics *teaching* rather than achievement. In other words, we center our discussion on the scale of *provision* of mathematics instruction, both in subject-matter terms and in terms of the number of students who are given opportunities to study the subject. We address the contextual and curricular variables that seem to be associated with the different scales of *provision* of instruction found in the various systems. Yield seen as *attainment* is explored in Volume 2 of this series.

Before embarking on our consideration of these questions it must be pointed out that there is a significant difference between FIMS and SIMS in their designs—and therefore significant differences in the kinds of questions about yield that can be asked between the two studies. In SIMS, in contrast to FIMS, there is only one target population at the end of secondary school. As a result, SIMS is not able to examine the opportunities to learn or the attainments of students *outside* the population of mathematics specialists included in Population B. We cannot, therefore, examine Postlethwaite's (1967) question, "How many are taken how far?" but must instead ask, "How many *mathematics specialists* are being produced by the various school systems being investigated?" and, "How much mathematics is being offered these specialists?" We assume that the classrooms in which Population B students are learning are the settings in which the lion's share of the teaching of advanced mathematics is taking place. We further assume that the vast majority of students who are candidates for further specialization in mathematics—whether in a university studies in mathematics itself or in fields like the physical sciences, engineering or business—are found in these classes. We are, therefore, considering how different systems go about the task of diffusing an understanding of advanced (school) mathematics through their professional, industrial, and commercial cultures.

In stating the themes of this chapter in this way, two fundamental qualifications of the findings to be reported here must be acknowledged. First it is, of course, an oversimplification of much more complex realities to assume that Population B students are the only students within the participating systems receiving instruction in "advanced mathematics," even as this was defined by the SIMS grid and item pool. As Postlethwaite (1967) shows, there were systems participating in FIMS in which students who were *not* mathematics-science specialists evidenced higher levels of attainment than did specialists in other systems. (See Figure 6.2.1.) This comes about because there can be widespread teaching of "advanced mathematics" prior to the Population B year—and in some systems many students are exposed to such teaching even though only a small proportion of the group will proceed to the more "advanced" mathematics of the terminal secondary year. Such teaching can, of course, create a large pool of students with some experience of, say, calculus and it may well be that such teaching provides a quite adequate gateway to many mathematics-using occupations and professions.

The kind of qualification implicit in this observation may be generalized. Widespread teaching of advanced mathematics can take place *before* the Population B year, *after* the Population B year, and *concurrently* with that year—in settings and schools that are not embraced within a national definition of "secondary school." Thus, in England and Wales, 26 percent of students at "O level" (year 10) take a mathematics course, in contrast to the 6 percent who take "A levels" (year 12) and who make up the Population B target group. A significant number of these students receive exposure to at least elementary advanced mathematics. In the United States (to take a different example) courses covering material included in the Population B grid are widely taught in third level (postsecondary) institutions and it is quite possible for students to enter such courses and to continue advanced studies in mathematics with backgrounds that do not extend past the content of grade 10 mathematics. Or to take yet another example, in many continental European school systems one or another course in mathematics is required of most or all students in academic secondary schools through to the terminal year. Many such courses include *some* of the content included within the SIMS Population B grid. In short, any estimate of the size of the pool of "mathematically-literate" students being produced by an educational system that is based on Population B enrollments alone is almost invariably an underestimate of the size of the "real" total pool of such students—although it would seem reasonable to claim that the bulk of those who will ultimately specialize in mathematics or mathematics-related fields found within a given system are nurtured by the main-line secondary school mathematics curriculum which culminates in the Population B course(s).

A second qualification of the findings to be considered here must also be emphasized. Systems differ markedly in the *constraints on curricular choice* in and around mathematics imposed on students. Thus, it is common in those systems with one or another form of course-based curriculum to treat mathematics as only one of a number of courses that students may elect at the upper secondary level. In England and Wales such freedom comes after the grade 10 level. In Finland *all* students are required to take substantial courses through to their terminal year although only one of the courses (the "long course") is regarded as a course appropriate for mathematics and science specialists. Needless to say, such curricular structures may have significant implications for retentivity.

6.3   Yield of Mathematics Education—Indicators

Retentivity, the proportion of the *age* cohort enrolling in Population B mathematics classes, can be interpreted as an index of the scale of teaching of advanced mathematics by a school system. Data on retentivity were presented in Chapter 3 and are recapitulated in Figure 6.3.1. While the modal system enrolls 10–12 percent of the age cohort in its Population B classes, there are clear, and at the high end of the scale, dramatic exceptions to this pattern.

As was seen in Chapters 4 and 5, systems also vary markedly in the scope of their *coverage* of the SIMS Population B "curriculum." To interpret the meaning of the retentivity data seen in Figure 6.3.1, we obviously need to calibrate levels of retentivity and the scope of a curriculum. It is an obvious speculation that high retentivity might be achieved within a particular system at the cost of a high level of coverage.

The implemented coverage data reported in Chapter 5 can be summarized in terms of three indices each of which defined a different aspect of the breath and depth of a system's coverage of the SIMS content domains:

1. Coverage of the SIMS *total item pool* (weighted by the emphasis given each content domain in the SIMS item pool). This index can be interpreted

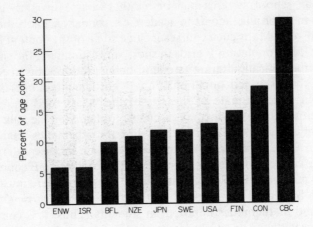

FIG 6.3.1   Population B: Retentivity of Mathematics Classrooms

as an indicator of the scope of a system's coverage of the *universe* of potential topics within advanced mathematics as this was defined within SIMS.

2. Coverage of the *Algebra-related topics* in the SIMS item pool, i.e., those items making up the SIMS content domains of Sets and Relations (7 items), Number Systems (19 items), and Algebra (25 items).

3. Coverage of *Elementary Functions and Calculus* as defined within the SIMS item pool (46 items).

Table 6.3.1 and Figure 6.3.2 recapitulate the findings reported in Chapter 5 for each of these content areas.

Table 6.3.1   *The Yield of Population B Mathematics*

Country	Total coverage	Algebra-related topics	Elementary functions/ calculus	Retentivity in Population B mathematics
Belgium (Flemish)	83	89	89	10
Canada (British Columbia)	51	79	35	30
Canada (Ontario)	68	73	83	19
England and Wales	77	79	88	6
Finland	85	92	88	15
Hungary	64	71	68	50
Israel	62	66	79	6
Japan	91	94	94	12
New Zealand	85	91	94	11
Sweden	80	87	88	12
United States	66	87	58	13
Mean	74	82	77	13
Median	73	83	85	12

Population B: Coverage

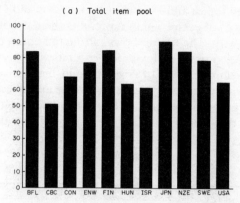

(a)   Total   item   pool

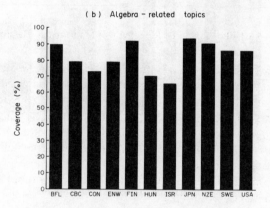

(b)   Algebra – related   topics

Coverage (%)

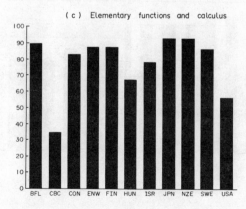

(c)   Elementary   functions   and   calculus

FIG 6.3.2   Population B: Coverage

Figure 6.3.3 is a representation of the yield of each system's mathematics instruction in the Population B year seen as the product of coverage in the Algebra-related topics and Elementary Functions and Calculus and retentivity. The height of each column represents the level of coverage and the breadth retentivity. The figure shows substantial differences in the yield of teaching in the various systems and exhibits a general tendency for a trade-off between the level of coverage and the size of the group being taught—with higher-coverage systems having lower levels of enrollment. However, a careful inspection of the figure also suggests that there are differences in the overall profile of yield associated with different content

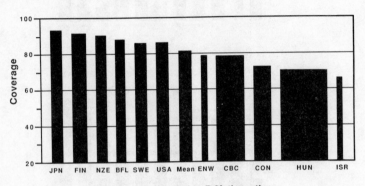

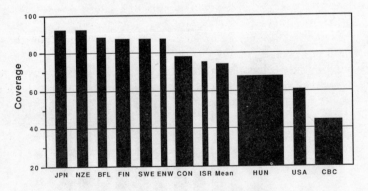

Fig 6.3.3   The Yields of Mathematics Instruction

areas, and also some systems with higher than expected enrollments given their level of coverage. Thus, the coverage profile (and, therefore, the yields) associated with the Algebra-related topics in the SIMS item pool is flatter than in the profile for Elementary Functions and Calculus. Canada (Ontario) has a higher level of coverage of Elementary Functions and Calculus than its total coverage would seem to suggest and a different overall level of yield in the Algebra-related areas than in Calculus. The United States, on the other hand, has a lower level of overall coverage than its retentivity level might suggest and a lower coverage of Elementary Functions and Calculus than its coverage of the Algebra-related topics would seem to foreshadow.

Table 6.3.2 summarizes the yields associated with each of the content areas reported in Table 6.3.1 in another way and provides a basis for clustering systems in terms of the overall character of their yields.

The vertical and horizontal axes on each figure represent the median of each dimension and divide the set of systems into "high" and "low" on each dimension—and into four quadrants each representing a different kind of yield. Thus, in Table 6.3.2(a), reporting coverage of the *total item pool*, six of the systems fall into the high coverage segment of the figure and five in the low coverage segment; five fall into the high retentivity segment, and seven in the low retentivity segment. When these dimensions are combined, Finland occupies the high-enrollment/high-coverage cell, Flemish Belgium, England and Wales, Japan, New Zealand and Sweden the low-enrollment/ high-coverage cell, Canada (British Columbia), Hungary, and Canada (Ontario) the high-enrollment/low-coverage cell and Israel and the United States the low-coverage/low-enrollment cell. This general pattern recurs in each of the other figures seen in Figure 6.3.2. But it is interesting to note the changing placements of some systems in the different content areas: thus, the United States moves from the low/low cell in Total Coverage and Elementary Functions and Calculus to the high/low cell in the figure for the Algebra-related topics; Canada (Ontario) moves from the high/low cell in the Total Coverage and Algebra-related topics assessments to the high/high cell in Elementary Functions and Calculus. But, while it is important to note that there are shifts across categories and assessments across the subject areas of Figure 6.3.2, it would seem that the overall placement of each system typically holds across the content areas. It would also seem that the more expansive curricula are related to lower levels of enrollment in Population B classes. This might suggest that such a curriculum is accessible, in the main, only by abler students. However, as we suggest this possibility, it becomes important to note that the enrollment levels *within* the cluster of systems incorporating a substantial component of Elementary Functions and Calculus in their curricula range from 6 percent in the cases of England and Wales and Israel to 15 percent in the case of Finland and 19 percent in the case of Canada (Ontario).

Table 6.3.2   *The Yields of Mathematics Education*

(a) Total Coverage

Retentivity

High	14	Low
Finland		Belgium (Flemish)
		England and Wales
		Japan
		New Zealand
		Sweden

Coverage 73

Canada (British Columbia)		Israel
Hungary		United States
Canada (Ontario)		

Low

(b) Algebra-Related Topics

Retentivity

High	14	Low
Canada (British Columbia)		Belgium (Flemish)
Finland		England and Wales
		Japan
		New Zealand
		Sweden
		United States

Coverage 83

Hungary		Israel
Canada (Ontario)		United States

Low

(c) Elementary Functions and Calculus

Retentivity

High	14	Low
Finland		Belgium (Flemish)
Canada (Ontario)		England and Wales
		Japan
		New Zealand
		Sweden

Coverage 85

Canada (British Columbia)		Israel
Hungary		United States

Low

What can the SIMS database tell us about the antecedents of a "decision" by a school system to adopt one or another of these yield-patterns? The answer to this question is, of course, embedded in the histories of the various school systems that participated in SIMS. Nevertheless, some empirical findings of the Study do throw some interesting light on the interaction between past and present.

6.4   The Context of Yield

Any consideration of findings about the yield of school mathematics must begin by contextualizing mathematics as a school subject in the structures and cultures of the school systems being examined. A large part of this context is, as we have suggested, embedded in history. Thus, the curriculum we have identified as containing a substantial component of the calculus emerged in Europe in the years before the First World War as an idea and possibility and became institutionalized in the school systems of Europe in the interwar years (Wojciechowska 1986). In contrast to this comparatively rapid implementation of this "new" curriculum in mathematics in Europe—and in the school systems of the European dependencies of that era that took their lead from developments at "home"—the North American systems, and particularly the United States' school system, did not implement that curriculum at that time. Indeed, the United States did not make a sustained attempt to introduce the calculus in its schools until the 1950s and 1960s, and, as was noted in Chapter 5, even at the time of SIMS only about one-fifth of the Population B mathematics classrooms in the United States offer a course in Elementary Functions and Calculus that is roughly equivalent in content to that available in most other SIMS countries (McKnight et al. 1987). Such differences between school systems in the reception of a calculus-based curriculum are, of course, related to larger differences in both the immediate contexts of secondary school mathematics and the social and cultural forces that play on schools. Thus, in contrast to other nations' university mathematics curricula in the interwar years, the American university did not offer calculus until after the first year. The movement of calculus down to the first year of university curricula took place well after the Second World War and was not firmly in place until the effects of the "New Mathematics" movement of the late 1950s and the 1960s were realized. This contrast between the American and European university is itself revealing in that it symbolizes what must be regarded as one of the core differences in the historical structure and culture of both schools and universities on the two sides of the Atlantic. In the United States the school and the university overlap in their curricula and structures, and the university assumes the task of both general and specialized education in the first 2 years of its 4-year baccalaureate program. In Europe, on the other hand, the institution of the university has long dominated the work of the traditional academic secondary school and specialized education of one or another kind is the norm even in the upper secondary school (Clark 1985). It was in such schools, with their concern for articulation between the work that might be done in universities and the work of the academic secondary school, that an introduction to calculus as the basis for later, more specialized studies in mathematics or physics and the like came to be seen as a proper end-goal of school mathematics programs. The schools in which such a curriculum was implemented were, of course, schools for a minority and saw their goals in terms of the "needs" of that minority. These schools were, moreover, largely for boys rather than girls and were closely linked to the cultural and occupational aspirations of advantaged social groups.

However, while this exploration of the history of the contexts in which the different curricula and patterns of yield associated with American and European schools and universities emerged is important, it does not address the changes that may have taken place in both kinds of school systems in more recent years. It does not address the interplay that has and does occur between pasts and presents. It does not address the characteristics of societies like Japan that fit uneasily into frameworks derived from European or American experience. Thus, comprehensivization and mass secondary education have emerged in Europe over the past two decades or so and, as ideals, have been given different forms in different European nations. Likewise, North American school systems and societies have changed in equally dramatic ways in the years since the Second World War. Thus, while we cannot ignore history, it is important to consider the present *contexts* of teaching in the various systems that we might be concerned with if we are to come to terms with the forces, and the meaning, of the kind of differences in yield that were pointed to above—and the prospects for increases in yield in different places.

Such explanation of the present is the task of the balance of this chapter. As we shall see we will have to come back to history as we consider our findings. Our immediate goal, however, is to discern the forces that might *now* be playing a role in determining one or another kind of yield for mathematics instruction. Thus, the next section will consider some of the between-system differences in the larger contexts of the school systems being discussed in this chapter to see what light these differences might throw on yield. The last section of the chapter will consider the curriculum itself as a factor that might play a role of reinforcing and creating context. Between these sections we will consider the particular issue of gender and its relationship with yield.

6.5   The Contexts of Population B Mathematics

There are a host of potential influences that could be reasonably explored with a view to understanding the forces that impinge on a school system to produce a level of yield in any subject. Here we will focus on only a small subset of the potential universe of such forces and indicators and consider three clusters of possible antecedents of the differences in yield discussed earlier.

1. The level of participation by the relevant age cohorts in formal education or training in both the Population B year (where pupils are typically 17 or 18) and in the third-level (tertiary) education.

2. The level of participation by girls in both upper secondary and third level education.

3. The degree of socioeconomic selectivity associated with school mathematics courses.

This choice of indicators, on the one hand, permits us to test the relationship between Population B enrollments and the pool of potential candidates for Population B classrooms and, on the other hand, explore some of the forces at work in any national system which determine entry to Population B programs. The primary focus of this discussion will be on determinants of retentivity in Population B.

Table 6.5.1 presents indices for each of these potential determinants of yield. It is interesting to note the wide between-system variation in almost all of the indicators. Thus, the percentage of the Population B-year age cohort enrolling in one or another form of educational or training program or institution ranges from 42 percent in the case of Sweden to 92 percent in the

Table 6.5.1 *The Contexts of Population B Mathematics*

	BFL	CBC	CON	ENW	FIN	HUN	ISR	JPN	NZE	SWE	USA
A. Participation Rates											
1. Proportion of Pop B age cohort in education and training—percent	65	82	49	56	59	50	60	92	40	42	82
2. Proportion of age group (20–40) in third-level institutions—percent	37	NA	NA	20	33	12	34	31	27	37	56
B. Female Enrollment Rates											
3. Participation in Pop B mathematics—percent female	41	41	37	34	43	63	43	22	36	26	44
4. Participation in Pop B level secondary grade—percent female	51	51	47	47	–	63	56	40	42	28	49
5. Participation in third level institutions—percent female	37	NA	NA	38	50	NA	48	23	42	37	50
C. Social Context											
6. Social selectivity of Pop B mathematics classes—increase from Pop A to Pop B in proportion of students reporting their father's occupation as "professional-managerial"	NA	43	101	175	123	NA	325	NA	162	127	91

case of Japan. Third level enrollment rates vary from 14 percent for Hungary to 56 percent for the United States. Female enrollment in the Population B grade of the schools from which each system's Population B sample was selected ranges from 37 percent for New Zealand to 63 percent for Hungary. The socioeconomic selectivity of the Population B classrooms sampled in different systems was defined in terms of the increase from Population A to Population B in the number of pupils reporting that their father's occupation was one classified by national committees as "professional or upper managerial." In the case of Canada (British Columbia), 29 percent of Population A pupils and 41 percent of Population B pupils were from such families, yielding a selectivity index of 43. In the case of Israel the increase was from 8 percent in Population A to 34 percent in Population B, yielding a selectivity index of 325.

Can we find any patterns of association between such forces in the context of Population B classes and the components of yield that we have been considering, level of coverage of the Population B item pool, and retentivity?

Coverage: Figure 6.5.1 presents a set of plots indicating the relationships between levels of coverage of the Algebra-related and Calculus items in the SIMS item pool and the contextual indices reported in Table 6.5.1. An inspection of these plots suggests little that might be regarded as unexpected in the light of our previous discussion of the differences between "North American" and "European" advanced mathematics curricula. As their histories would foreshadow, two of the North American systems (Canada (British Columbia) and the United States), are more inclusive in their overall school enrollment rates than are all other systems except Japan. But, interestingly, these systems differ dramatically in their levels of enrollment in Population B mathematics. It is also interesting to note that Hungary and Israel are clearly separated from the other systems— but, as has been noted several times, Hungary is unique (within the SIMS systems) in requiring all students in the terminal secondary year to take a course which could be seen as meeting the international definition of Population B courses. Israel is the only middle-income nation in the set of systems being considered here.

There is, furthermore, considerable variation in the contexts associated with high-coverage curricula. Thus, while such systems do have a narrower band of enrollment levels in Population B than do the low coverage systems, the range is still very substantial. *It would seem that the high-coverage curriculum is, when viewed across systems, relatively robust in that it can be taught to comparatively large numbers of pupils in at least some contexts.* This would seem to have significant implications as we consider the yields of mathematics instruction which *might* be attained within school systems.

Retentivity in Population B: Although there are some significant differences in the levels of coverage attempted within the different systems, an inspection of Figure 6.3.3 (see above) suggests that it is varying levels of retentivity

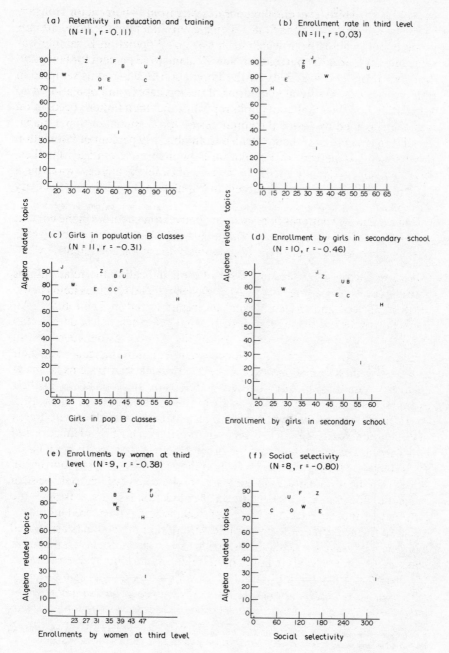

Fɪɢ 6.5.1.A   Correlates of Coverage – Algebra related topics

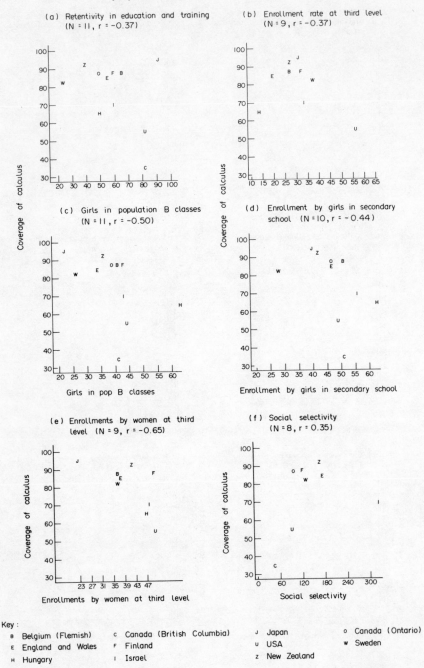

FIG 6.5.1.B   Correlates of Coverage – Elementary Functions and Calculus

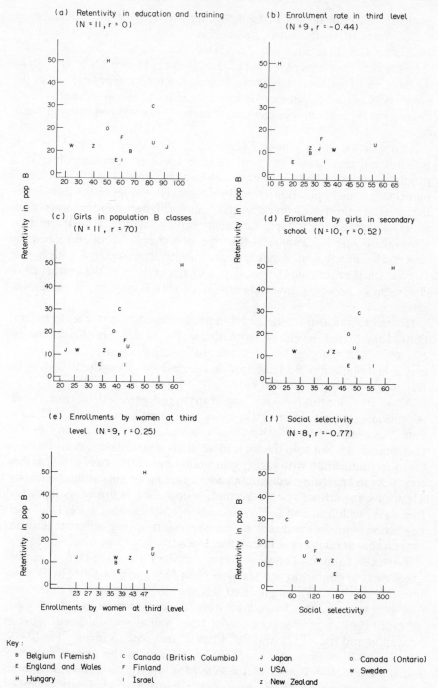

FIG 6.5.2 Correlates of Retentivity

particularly within the set of high-coverage systems. What differences in the contexts of mathematics instruction at the Population B level are associated with these differences in retentivity?

Figure 6.5.2 presents plots that parallel those in Figure 6.5.1. They show the patterns of association between Population B retentivity defined as the proportion of the Population B age cohort enrolling in advanced, i.e., Population B, mathematics and the contextual indicators reported in Table 6.5.1.

An inspection of the plots in Figure 6.5.2 would suggest that educational policies, and milieus, which support higher levels of enrollment by young people, and particularly women and girls, have consequences for levels of participation in Population B mathematics. Thus, Figure 6.5.2 indicates that overall enrollment levels in third level (higher) education are positively associated with high retentivity in Population B mathematics. Likewise, Figure 6.5.2 suggests that overall higher levels of enrollment by girls in the Population B grade is also positively associated with higher Population B enrollment levels. The social selectivity of the mathematics classroom is, however, inversely associated with Population B enrollment rates.

However, it is also interesting to note, in the context of these observations, that higher levels of participation by 17–18-year-olds in formal education and training do *not* appear to have any clear effect on enrollments in Population B. How can this seemingly inconsistent finding be reconciled with our other findings?

As we consider this issue it is important to recognize a definitional issue: educational systems differ markedly in the extent to which they differentiate "secondary" from other forms of education or training in the postcompulsory years, in the ways in which they articulate various forms of secondary education with higher education, and in the ways in which they treat full- and part-time education and/or training within national statistics. Thus, in some school systems secondary education is firmly distinguished from, say, technical and further education while in other systems no such distinction is made—and such differences result in very different statistics for enrollment rates in the Population B year.

Table 6.5.1 reported enrollment rates at the Population B level as total enrollment in education and training, both full- and part-time, by Population B age youth. We emphasized this statistic on the ground that it was less contaminated by what are, from a cross-cultural point of view, arbitrary distinctions between school types and also distinctions between education and training. However, in all of the systems involved in SIMS, Population B mathematics was defined in terms of the *secondary* school and the Population B course(s) selected for investigation were predominantly preuniversity in their orientation. This definition clearly delimited Population B to a particular set of courses and contexts, although it is important to note that most systems reported that there would be few, if any, pupils of Population B-age taking courses similar in character to their Population B courses outside their target populations.

These definitional issues have important consequences as we seek to interpret the relationship between overall school enrollments and retentivity. Thus, in England and Wales 21 percent of the 17-year-old cohort attend "schools," 10 percent attend institutions offering "nonadvanced further education" on a full-time basis and another 24 percent enroll in nonadvanced programs on a part-time basis, giving a total 17-year-old enrollment rate of 56 percent. However, the pool of candidates for Population B mathematics classes comes from only the 21 percent enrolling in "schools." In the United States, on the other hand, 82 percent of 17-year-olds are enrolled in a single institution, the high school; while some students in these schools will be enrolled in what is, in effect, a part-time basis and others will be taking occupational programs, *all* students in these schools constitute potential members at least of the Population B group. (See Table 3.7.1).

Such differences in the *structures* of postcompulsory education reflect, of course, very different social and educational assumptions; at the one pole all forms of education and training are housed within one institution and all forms of such education and training are seen as equal in formal standing—with the implication that all offerings of the institution are seen as, in principle, equally available to all. On the other hand, the various forms of education and training are housed in separate institutions, and, typically, complex social and education gradations are made between institutional types and the programs they house.

Given the variation in the forms of delivery of postcompulsory education and training found among the systems participating in SIMS, it might be predicted that the levels of overall enrollment of Population B-age students are not related to Population B enrollment rates. As seen in Figure 6.5.2, this is indeed the case. But another question emerges: What is the relationship between enrollments in "academic" secondary education as this is defined in each system and enrollments in Population B? Are policies which have the effect of incorporating greater and lesser proportions of the age cohort within a single form of postcompulsory school associated with higher levels of enrollment by students in advanced mathematics?

Figure 6.5.3 plots the relationships between the proportion of the age cohort enrolling in the school-type that participating system designated as its Population B target school-type and retentivity to Population B. Two clear clusters would seem to be present in the plot. There is a cluster of lower-enrollment systems in which it seems that the higher levels of enrollment in the secondary school are associated with higher levels of retentivity to advanced mathematics. However, in the cluster of higher enrollment systems this is less clearly the case and it would seem that there may be

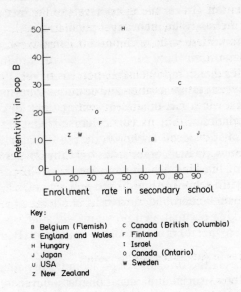

Key:

B Belgium (Flemish)   c Canada (British Columbia)
E England and Wales   F Finland
H Hungary   I Israel
J Japan   o Canada (Ontario)
U USA   w Sweden
z New Zealand

FIG 6.5.3   Participation Rates in Secondary Schools and Retentivity

framing factors at work in these systems that inhibit enrollment in advanced mathematics which are different from those in the lower enrollment systems.

What might these frames be? We cannot answer this question directly but Figure 6.5.2 suggests some of the out-of-school forces that appear to be associated with enrollment levels in advanced mathematics in at least some systems. In the case of Israel, socioeconomic selectivity would seem to be tightly linked with enrollment in advanced mathematics. In the case of Japan, gender appears to be similarly associated with enrollment in advanced mathematics. It is interesting to note that we cannot account for the low retentivity of advanced mathematics in the United States by invoking forces of these kinds.

In summary then the plots presented in Figures 6.5.1 to 6.5.3 seem to suggest an interpretable cluster of forces playing on the mathematics curriculum to determine the yield of advanced mathematics instruction found within a system. These forces are, moreover, seemingly consistent across most systems. Thus, it would seem that the ways in which schooling is distributed within a system play a major role in determining the accessibility of mathematics instruction. Furthermore, policy trends that support the expansion of a common, comprehensive education within a single school type do appear to increase opportunities to enter the academic core of the traditional secondary school. The major role that such policies would seem to play in this process are apparently associated with the support they offer to the enrollment of girls in the upper levels of the secondary school, and once in school, their enrollment in advanced mathematics. Such policies also seem to be associated with declining socioeconomic selectivity in the mathematics classroom.

We might infer that comprehensivization has the effect of decreasing the association between status variables and enrollment in school-types and that the presence of a wider mix of students within one comprehensive school-type, which includes within its curriculum introductory mathematics as distinct from, say advanced arithmetic, plays a major role in increasing opportunities for individuals. Significantly, in those systems which have had a long tradition of teaching a high-coverage curriculum, such policies and directions within the school system do *not* seem to have the effect of increasing levels of participation in advanced mathematics at the expense of coverage.

6.6 Gender-bias in Advanced Mathematics

Table 6.5.1 shows substantial variation in the levels of participation by girls in Population B classes across the systems considered in that Table. Figures 6.5.1 and 6.5.2 suggest that one important determinant of these various retentivity levels would seem to be the overall enrollment levels of girls in the schools from which the Population B sample was drawn and that higher levels of participation by women and girls in school have important consequences for yield. In that discussion the focus was, of course, yield and the contextual conditions which seem to support higher levels of yield from mathematics instruction. From that perspective girls become only one group whose opportunities are enhanced by policies which support the development of common comprehensive forms of provision of upper secondary education.

However, the widely-perceived gender bias of mathematics merits further discussion. In this section we will revisit some of the findings of the previous section and consider the issue of female enrollment more directly using the full set of data available to the Study. Thus, in addition to the set of systems that constitute the principal focus of this chapter, data on enrollment by girls in Population B mathematics are also available from French Belgium, Hong Kong, Scotland, and Thailand. Furthermore, data on changes over time in the levels of girls' participation in advanced mathematics classes are also available from FIMS for a subset of these systems.

The issues that justify attention to this aspect of the yield of mathematics instruction have been a major concern in many societies in recent years. The patterns of gender bias that have marked mathematics instruction in many school systems have had important consequences for the opportunities available to women and girls and have been associated with forms of occupational segregation which, in a climate of concern for equity, have seemed unsustainable in principle and practice. However, a basic question would seem to circle around this issue as it bears on both enrollment and achievement: What are the causes of the seeming problems of girls with, and the seeming lack of interest of girls in, mathematics? A host of factors ranging from gender-linked reasoning and problem-solving abilities to genetic differences in brain organization have been invoked to account for the problem. But there would seem to be an emerging consensus that gender differences in both patterns of achievement in mathematics and enrollments arise instead from pervasive psychosocial factors. However, as Harnisch (1984) and Harnisch et al. (1986) note, all such attempts at explanation have been plagued by both methodological weaknesses and, more importantly, the limitations of work that is undertaken within single cultural communities. As Harnisch et al. (1986) note in the introduction to their study of mathematics achievement of the FIMS samples, "earlier work with English-speaking samples leads us to expect that 17-year-old boys in the present sample will score higher on tests of achievement and exhibit more positive attitudes toward mathematics than do girls. But are the differences universal across cultures?" Their study, using the FIMS database, found that gender differences in mathematics attitudes and achievement were pervasive across cultures and usually favored males. However, they noted that there were indications in their data that the differences between sexes that they found were not immutable and "were malleable in the context of the family and school."

The database investigated by Harnish et al. (1986) was collected in the 1960s and the years since FIMS have seen major public concern with gender bias as a larger issue in many societies. Such changes in the *context* of school mathematics have had significant effects on female participation in mathematics-dependent occupations in at least some societies. We can only wonder what these such might also suggest about the mutability and malleability of gender as a force in secondary school mathematics.

Figure 6.6.1 summarizes the findings on enrollments by girls for *all* systems which participated in the Population B phase of the Study. The striking finding which emerges in the Figure is, of course, the substantial between-system variation in the rates of female participation at the extremes of the distribution. Thus, the overall ratio of girls to boys in these classes ranges from more than three boys to every one girl in Japan to one boy to one girl in Thailand— and in the special case of Hungary (where all students in the terminal grade of the secondary school were included in Population B), there were more girls than boys in Population B classes. It is clear though, that systems like Japan and Sweden at the lower end of the distribution and Thailand and Hungary at the high end are special cases. In most systems the ratio of girls to boys in Population B mathematics was about 2 or 3/1.

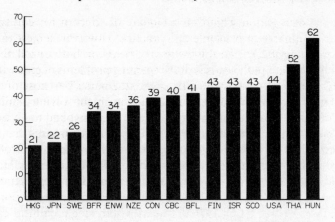

FIG 6.6.1   Enrollments by Girls in Population B Classrooms

This general pattern is very different from the pattern of the middle 1960s. Figure 6.6.2 presents a comparison of the rates of female enrollment in Population 3a of FIMS and Population B from SIMS in the systems in which the two populations can be regarded as similar in character. The figure suggests that a substantial "improvement" in gender equity has occurred in most of these systems over the two decades that separate these studies. But, interestingly, it is also noteworthy that in Japan there has been a *decline* in the relative level of female participation in advanced mathematics and virtually no change in Sweden. However, overall, these differences over this 20-year-time span in the level of girls' participation in advanced mathematics seem to suggest that female enrollment in mathematics, if not achievement, should be thought of as malleable and mutable.

We cannot attempt a specific exploration here of the range of psychosocial conditions, and the changes in those conditions, that might explain the

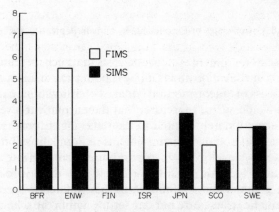

FIG 6.6.2   Changes in Male/Female Enrollment Ratios in Population B:
FIMS to SIMS

different patterns seen in Figures 6.6.1 and 6.6.2. However, we can attempt to see what a preliminary cross-sectional analysis might suggest about the correlates of varying levels of enrollment by girls in Population B mathematics. Figure 6.6.3 plots the relationships between the indices presented in Table 6.5.1 indicating the overall levels of female participation in the Population B-grade and third-level schooling in the various systems. It seems clear that enrollment by girls in advanced mathematics classes is a manifestation of the *general* level of participation by women and girls in upper secondary and third level education. Figure 6.6.4 which plots the relationship between female participation in Population B classes and overall levels of retention in education and training of Population B age

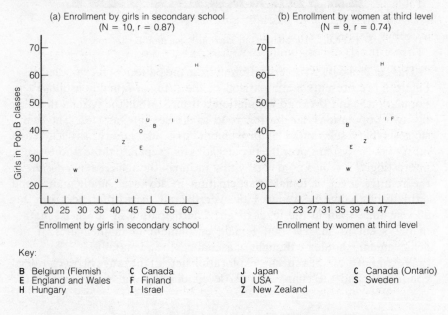

FIG 6.6.3   Correlates of Enrollments by Girls in Population B Classes

youth in an augmented set of systems suggests that such increased participation by females in education and training is both a cause and a consequence of a larger process of "incorporation" of age cohorts within the school as an institution. It would seem, therefore, that the enrollment levels of girls in mathematics are intimately related to the same set of forces we discerned earlier as we considered overall enrollments in Population B mathematics: education systems "develop" by incorporating age cohorts into the school as a general and generic institution and in this process, status variables like gender tend to diminish in their influence on school attendance. Mathematics classes are but one setting within the school and, as the process of incorporation proceeds, girls, as well as pupils from disadvantaged social groups, take their place in those classrooms alongside their male, socially-advantaged peers. We might suspect that as this larger process occurs the differences that hitherto differentiated the achievement of girls and boys also fade. The significant exception to this process as it is seen in SIMS is, of course, Japan. There it would seem that gender as a status variable is particularly significant and overrides the general process seen in most, if not all, other systems.

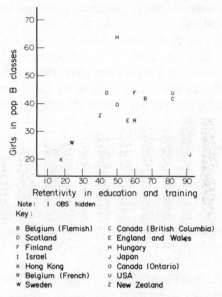

Note :   ı   OBS hidden

Key :

B	Belgium (Flemish)	c Canada (British Columbia)
D	Scotland	E England and Wales
F	Finland	H Hungary
I	Israel	J Japan
K	Hong Kong	o Canada (Ontario)
R	Belgium (French)	u USA
W	Sweden	z New Zealand

FIG 6.6.4   Enrollments by Girls in Population B Classes and Retentivity in Education and Training

However, at the same time as we make this observation it would also seem important to note that in virtually all of the societies that participated in SIMS, gender has not totally disappeared as a variable determining the level of participation in mathematics. Thus, it may be that there are specific gender-linked traits which interact with instructional practices that do influence the attitudes and achievement, and therefore enrollment, of girls in mathematics, but it seems more reasonable in the light of the data presented here—and experience—to suggest that this is not the case. It is more likely, we believe, that the inhibiting factor seen in virtually all of the societies considered here is associated with the less than complete penetration of the institutionalization of universalism as a value within a society.

6.7   Yield and Pace

To this point this chapter has focused on the mathematics curriculum cross-sectionally. We considered whether the character of the Population B curriculum seemed to influence retentivity to any significant degree and concluded that it did not seem to be a significant factor in limiting access to advanced mathematics for at least 20 percent or so of an age cohort. Above that point the findings from SIMS are unclear. Systems that enroll more than 20 percent or so of their age cohorts do cover a less demanding curriculum—but there are only four such cases. In systems enrolling under 20 percent of their age cohorts in mathematics variation in retentivity seemed to be associated with larger social and institutional forces and not the curriculum as such.

However, such a conclusion cannot terminate our concern for the curriculum as a potentially significant factor determining enrollment rates in advanced mathematics and, therefore, yield. Both sociologists and scholars of the curriculum are increasingly recognizing that it is the curriculum of the school that is the actual agency that determines the ways in which larger social forces penetrate the work of the school to create the sorting and selection that gives a particular school system its special character. The forces within schools and subjects that determine these processes must, of course, be seen as being built into the situation. But to be "effective" these forces must be invisible, part of what Bourdieu and Passeron (1977) term the facade of "neutrality" that schooling has in a society and which provides a cover for the "symbolic violence" that forces certain values on teachers and students. To put this another way, the practices that mark a subject and serve to give that subject its role, place, and function in the school must be accepted by publics and teachers if they are to do their work—but this invisibility makes them the harder to evaluate. One function of comparative cross-national studies like SIMS is to open such "invisible forces" for questioning.

What are the unrecognized forces within mathematics curricula that seem to be associated with, for example, the levels of yield that are seen within systems? *Mathematics Counts* (1982) draws attention to one candidate insofar as England and Wales is concerned. In its discussion of the examination courses that English students take in the upper secondary school, the report notes that:

Very many pupils in secondary schools are being required to follow mathematics syllabuses whose content is too great and which are not suited to their level of attainment. Efforts to introduce pupils to as much of the examination syllabus as possible result in attempts to cover the ground too fast for understanding to develop. The result is that very many pupils neither develop a confident approach to their use of mathematics nor achieve mastery of those parts of the syllabus which should be within their capacity (pp. 132–33).

Implicitly, if not explicitly, this argument draws attention to the place of *instructional pace* as a variable influencing achievement. When the pace of coverage which is embedded in a course of study is too fast for a target group of pupils, effective teaching become problematic. *Mathematics Counts* argues that this is the case for many pupils in English schools, and that this has come about because those curricula have not changed in their scope for

40 or more years *despite major changes in the aptitudes of students that have come about as a result of the emergence in England of mass secondary education.* Assumptions about coverage that were derived from a world very different from the present world of English schools have held fast and, as a result, mathematics has become a very restricted gateway for many students. One outcome of the sorting that is the product of teaching within such curricula is perhaps the preservation of sociocultural character of the "old" mathematics classroom in the "new" school, but a more important outcome is substantial attrition in the numbers of students who could be learning advanced mathematics if selection were less demanding.

In the context of this argument it is interesting to note that despite substantial increases in enrollment in upper secondary education, there has been virtually no change in the proportion of the age group enrolling in England and Wales' A-level mathematics courses between FIMS and SIMS. At the time of FIMS, 5 percent of the age cohort was enrolled in courses that were the equivalent of SIMS Population B courses (Husén 1967); at the time of SIMS, 6 percent of the age cohort enrolled in Population B courses. But between these years enrollments in the terminal year of the English secondary school system rose from 12 percent of the age cohort to 21 percent. In contrast, Population B enrollments in Finland increased from 7 to 15 percent of the age cohort as enrollments in the Population B school-type have increased from 14 to 35 percent. In Japan, enrollments in Population B increased from 8 percent of the age cohort at the time of FIMS to 12 percent at the time of SIMS; overall school enrollment in Japan increased from 57 percent of the age cohort to 92 percent (Husén 1967). *Such differences in the responsiveness of participation rates in advanced mathematics to increases in overall school enrollments have obvious implications for yield.*

An impressive body of research supports the attention that *Mathematics Counts* gives to *curriculum pace* as a critical variable influencing the success that students have with any school subject. We can also speculate that this problem of pace may be even more important in the junior secondary years than it is in the middle or later grades. As we noted earlier, it is in these grades that the most dramatic changes in mathematics instruction have taken place as comprehensivization has brought about the teaching of mathematics to all students, at least in these years, and in many systems it is in the junior secondary years that the fundamental sorting of students takes place as students are streamed into courses and/or tracks only *some* of which articulate with advanced school mathematics. (See, for example, Ball 1981 and Evans 1985.)

The kind of analysis of a curriculum we have been foreshadowing here is historical and the exploration of its implications in particular systems or sets of systems is inevitably outside the scope of a study like SIMS. But the hypotheses are so important for the issue of yield that it becomes interesting to see if the arguments of the Cockcroft Report can be extended beyond the borders of one school system. Fortunately, one curricular investigation undertaken within SIMS provides a data set for such an investigation.

In 1959, as part of the background studies for the Organization for European Economic Cooperation (OEEC, now OECD) seminar *New Thinking in School Mathematics* (OEEC, 1961), national educational systems were asked to indicate on a test-like questionnaire the year and grade in which the content underlying each item was introduced to pupils. In 1980, a slightly revised version of the OEEC Questionnaire was prepared for SIMS and was sent to each participating country. The *Revised Questionnaire* (RQ) contained 39 items that sampled upper elementary and secondary school mathematics. The responses to this instrument reflected, of course, the *intended* curriculum and in secondary schools were targeted on the curriculum of the university-bound track or stream.

This Revised Questionnaire is set out in Figure 6.7.1 and the data it yielded are presented in Hirstein (1980). Table 6.7.1 reports the mean year in school in which all items and the Algebra, Geometry and Calculus item clusters are introduced in each of the systems being considered in this chapter.

Table 6.7.1   *Pace: Mean Year in School of Introduction of Various Mathematical Contents*

	All Topics	Algebra	Geometry	Analysis
Belgium (Flemish)	9	9.5	8.1	11.3
Canada (British Columbia)	8	10.2	9.5	12.0
Canada (Ontario)	10	11.8	9.3	13.0
England and Wales	9	8.8	7.5	10.3
Finland	10	9.7	7.2	11.0
Hungary	7	8.0	7.5	11.3
Israel	1	9.5	8.3	11.3
Japan	9	9.2	7.5	11.0
New Zealand	9	7.8	7.6	11.0
Sweden	10	9.9	8.8	11.0
United States	9	9.0	8.0	12.0

As seen in Table 6.7.1, there is considerable between-system variation in how the content embedded in the RQ is deployed across the grades in the various systems. The mean grade of introduction of All Items ranges from 7.0 in the case of Hungary to 10.0 in the cases of Finland and Sweden. For Algebra, the means range from grade 7.0 (Finland) to grade 9.5 (England and Wales) to grade 12 in Canada (British Columbia), Canada (Ontario) and the United States.

Mean grade of introduction of content is not, of course, an indicator of pace of coverage within a grade, the issue that concerned the *Mathematics Counts*. But it is perhaps an indicator of the relative pace associated with an "overall curriculum." Thus, we assume that those systems which report a mean grade of introduction of, say, Algebra, which is below the mean or median grade for all systems, expect their younger students to move faster than most systems. Likewise, it would seem as if those systems that report

Items marked with an asterisk (*) were not included on the Royaumont Questionnaire.

In which school year and student age does each of the following problems **FIRST APPEAR** in the mathematics curriculum? Answer separately for each appropriate student category.

A. ARITHMETIC

1. $68 + 25$

2. $804 - 347$

3. The multiplication tables to 10×10 or beyond

4. $784.92 +$
    27.38
    63.67
    591.59
    $\overline{}$

5. $2\frac{3}{5} + 5\frac{7}{12}$

6. 684×342

7. $2\frac{3}{4} \div 1\frac{5}{8}$

8. 375.24 divided by 17.3

9. Find mentally (use no paper and pencil) 4×239

10. What is the number of which $15\% = 6$?

*11. $12^2 \div 6^3$

*12. Find, by estimating, the decade in which $\sqrt{600}$ lies.

B. MATHEMATICS

*1. $(+ 10) - (- 25)$

2. Solve $3x - 7 = 2x + 4$

3. Solve $3x^2 - 15x + 18 = 0$

4. Solve $3x - y = 5$
    $x + 2y = 11$

5. Two trains each cover a run of 960 km. The one train takes 4 hours longer and averages 20 km per hour less than the other. Find the rate of each train.

6. Solve and discuss the solutions for various values of m:
    $(m - 1)x^2 + (2m + 1)x + (m - 2) = 0$

7. 37 is expressed in decimal notation. Re-write it in a system to base 6.

*8. Show for all sets A and B that
    $A = (A \cap B) \cup (A \cap B')$

9. Plot the graph for:
    $y = 3x + 2$

10. Plot the graph for:
    $y = \dfrac{3x + 5}{4x - 5}$

*11. f and g are functions such that
    $f(x) = x^2 + 1$ and $g(x) = x - 2$
    If $a = f(1)$, find $g(a)$.

12. Expand $(3x - 2y)^8$

13. Prove $1 + 4 + 9 \ldots + n^2 =$
    $\frac{1}{6}(2n + 1)(n + 1)(n)$
    using mathematical induction.

*14. Write $\dfrac{1}{2i + 3}$
    in the form $a + bi$.

15. Complete and prove for any angles α and β
    $\cos(\alpha - \beta) =$

16. Derive the law of cosines for any triangle.

Figure 6.7.1   *The Revised Questionnaire*

17. Calculate the area of a triangle given a base of 8 cm and an altitude of 5 cm.

18. Find the volume of a pyramid, the base of which is 16 sq. cm. and the height 12 cm.

19. Calculate the side of a right angled triangle given that the other side is 5 units and the hypotenuse is 7 units.

*20. $\overline{CA} = \overline{CB}$ and $\overline{AB} = \overline{AD}$

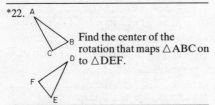

if m $\angle ACD = 30°$, find m $\angle BAD$

21. State and prove the Pythagorean Theorem.

*22.   Find the center of the rotation that maps $\triangle ABC$ on to $\triangle DEF$.

23. Find the derivative of the function
$y = 3x^2 - 5x$

24. Find the maximum and minimum values of
$y = x^3 - 12x + 5$

*25. Find $\int_1^2 (x^3 - 4x + 3)dx$

26. Find the period of

$y = 2.6 \sin \dfrac{\pi t}{60}$

27. Given two free non-zero vectors

AB and CD

i) find AB + CD
ii) find the scalar product

AB × CD

28. Assuming the associative, commutative, and distributive and cancellation laws, and the properties of 1 and 0, prove for the domain of positive and negative real numbers that
$(-x)(-y) = xy$

29. Prove: If a straight line is perpendicular to each of two intersecting lines, it is perpendicular to the plane of these two lines.

30. Find the equation of the straight line determined by the points
$A = (3,2)$ and $B = (4,½)$

*31. Find the center and radius of the circle whose equation is
$x^2 + y^2 - 4x + 2y - 4 = 0$

32. Determine the roots of $3x^5 - 7x^2 = 0$

33. What is the probability of getting at least 3 heads in a throw of 5 coins?

*34. How many different linear arrangements can be made with the following cards?
$\boxed{1}\ \boxed{1}\ \boxed{1}\ \boxed{2}\ \boxed{3}\ \boxed{4}$

35. Solve the inequation
$3x + 2 > 8$

36. Solve the system $y > ½x - 1$
$y < 3 - ⅓x$
$x > 0$

37. Draw the graph of $y = |x| - 2$
for the interval $-5 \leq x \leq 5$

Note: $|x|$ denotes absolute value of x.

38. Find the greatest common divisor of 42 and 5610.

39. Assuming a normal distribution with given mean and standard deviation, what is the probability in a sample of 10 of getting at least two individuals with a deviation of 2 or more standard deviations?

Figure 6.7.1   *The Revised Questionnaire* (continued)

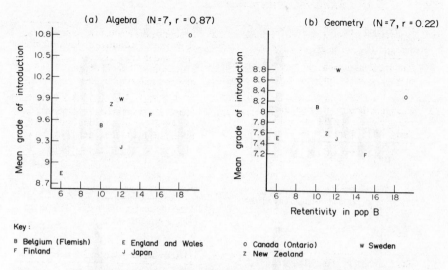

Key :

B Belgium (Flemish)   E England and Wales   o Canada (Ontario)   w Sweden
F Finland   J Japan   z New Zealand

FIG 6.7.2   Retentivity in Population B and Curricular Pace

introducing Algebra in a higher grade than that found in most systems are
moving at a slower pace. An extension of the argument of *Mathematics
Counts* would suggest that such differences may well affect both the quality
of learning and attitudes, and therefore, persistence with mathematics after
the point at which the subject becomes elective, i.e., retentivity.

Figure 6.7.2 plots the relationships between the mean grade of introduction of items in algebra and geometry and retentivity for the set of high-
coverage systems. It would seem as if, in the case of Algebra, there is a
suggestive association between curriculum pace in this sense and the ultimate retentivity of the school mathematics program. This association does
not emerge so clearly in the content area of Geometry.

But while the patterns suggested by Figure 6.7.2 are suggestive of interactions between the ways in which systems deploy their coverage across years
and the levels of ultimate attrition from mathematics, mean year of introduction is only loosely associated with the linkage between curriculum pace as a
frame within which teaching must be undertaken, the resulting quality of
teaching and learning and, ultimately, attrition (and therefore yield). *It is the
ways in which the total body of coverage is deployed across grades that is most
central to this argument.* What do the responses to the RQ suggest about
between-system differences in pace seen in this way, and the possible relationships of varying pace with attrition from school mathematics?

Figure 6.7.3 seeks to capture pace as described in the preceding para-
graph. The columns indicate the number of items from the RQ that a system
reported introducing in each grade and the height of the column is, by
implication, an indicator of the pace of the curriculum in that year. Each
figure permits comparison of the pace implicit in each grade within each

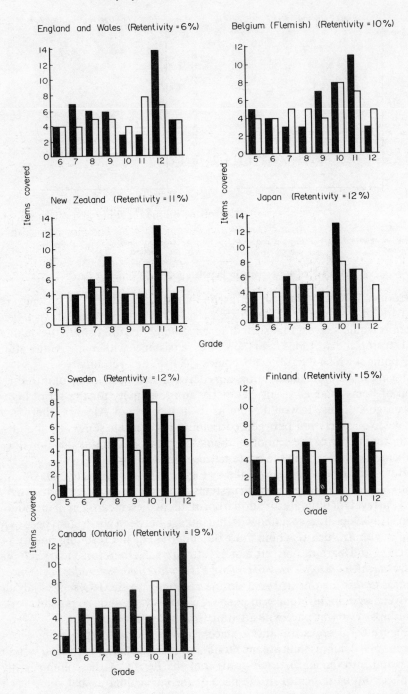

Fɪɢ 6.7.3   High-coverage Systems: The Pace of Curricula

system's curriculum while a comparison across profiles lets us compare the loads associated with particular grades in different systems. To facilitate such comparisons each figure also includes columns representing the median pace for each grade. Figure 6.7.3 includes only high coverage systems and the medians (indicated by the white bars) are those associated with the high-coverage systems.

An inspection of the profiles presented in Figure 6.7.3 makes it clear that no single pattern of curriculum organization is found across all of these systems. We can see variation in the extent to which curricula are marked by dis-junctions in pace across different years and, within a framework of such disjunctions, by differences in the absolute level of demand made on students in a single year. Thus, England and Wales has a curriculum which, in the context of the systems seen in Figure 6.7.3, is marked by an early fast-paced segment which is followed by two "lighter" years in grades 9 and 10. Japan shows a similar structure beginning in grade 7, although there is seeming decline in pace associated with the following years in contrast to England and Wales. Finland and New Zealand have a fast-paced year in grade 8 but it is interesting that, in terms of the number of items being covered, the speeded-ness of New Zealand's spiked year is greater than that of Finland. Canada (Ontario) begins with period of steady pace in grade 6 to 8 and follows this with a faster-paced grade 9 year, but the absolute increase in speededness of this year is less than that of New Zealand. Sweden moves much more quickly in year 7 than it did in the previous year and again in grade 9, but the pattern is one of overall movement toward a single year of peak load of students; the Flemish Belgian profile moves slowly upwards in load in grades 7 and 8 and then shows a sharp speeding of pace to a peak load in grade 11.

It is tempting to suggest that there might be a direct relationship between these profiles and the dispositions toward mathematics that are created in students, and so retentivity. And, it would seem that we can associate the kinds of curricular structures seen in Figure 6.7.3 with the ultimate yields of systems: England and Wales would seem to have a curricular structure that makes greater demands on students than do the curricula of such middle-yield systems as Flemish Belgium, Sweden and New Zealand—and even Japan—and the curricula of these systems, in their turn, would seem to make greater demands on their younger students than do the curricula of Finland and Canada (Ontario). Likewise (and, as we suggested above, this is also central to the issue of yield), England and Wales would seem to have curricular profiles which would be associated with lower rates of enrollment in an academic track than would the profile of, for example, Finland, to name one system which maintains a form of overt organizational differentiation of school- or curriculum-types at the upper secondary level. As we will see, it is one test of these speculations that they can be usefully applied to the subset of lower-coverage systems to yield similar interpretations of the interaction between curricular structure and ultimate retentivity.

In this discussion we have distinguished the high-coverage from the low-coverage systems as a way of acknowledging that the SIMS systems do not all reach the same end point in their coverage of mathematics in school—and this difference is, presumably, reflected in the overall curriculum in the secondary grades. The low-coverage systems represent, moreover, a very mixed bag of types of school systems and contexts. Levels of retentivity range from 6 percent in the case of Israel to Hungary's 50 percent. Israel is a middle-income society, whereas Canada (British Columbia) and the United States are high-income societies. Hungary is a society with a centrally-planned economy, whereas Canada (British Columbia) and the United States are market-oriented. Hungary enrolls *all* of its Population B age students in an advanced mathematics class, whereas in the other cases enrollment in advanced mathematics is elective—by contrast, in Canada (British Columbia) approximately 40 percent of the grade cohort enrolls in Population B, whereas in Israel 6 percent of and in the United States 13 percent of the grade cohort enroll in these courses. But this heterogeneity is matched by an almost equal heterogeneity of curricular structures and the associations between these structures and yields would seem to be interestingly tight.

Figure 6.7.4 presents the profiles of coverage across grades that emerged from the responses of the low coverage systems to the RQ. Canada (British Columbia) shows a flat, "undemanding" structure with few disjunctions in rates of coverage (pace) between grades before grade 12 when there is a rapid increase in pace. Israel and the United States, on the other hand, show a series of marked differences in the pace associated with various years both in the junior and middle secondary years as well as much more demanding absolute levels of coverage than those seen in Canada (British Columbia). Hungary has a profile which appears to be quite different from that of any other case we have considered in that the curriculum makes very substantial demands on grade 6 students, moves relatively slowly in grades 7 and 8 and then speeds up markedly in grade 9 and again in grade 10.

Although there may be a deceptive quality to this clarity, the substantial differences between the curricula and the outcomes of these low coverage systems would seem to offer some persuasive support to the argument we have been developing here. And with a small number of cases it is possible to consider also observations and data from other sources. For example, the *International Handbook of Education Systems* (Holmes 1983) reports that in Israeli academic secondary schools, "Failure rates in some subjects are high, particularly mathematics." We noted earlier that the socioeconomic selectivity of Israeli Population B classes is very high relative to other systems. In the United States an important problem that the SIMS national report (McKnight et al. 1987) raised was the tracking associated with the Population A (grade 8) year and the resulting tendency for lower track students to be directed out of the main-line college sequence of courses. Again, such practices would seem to flow almost necessarily from the structure of the

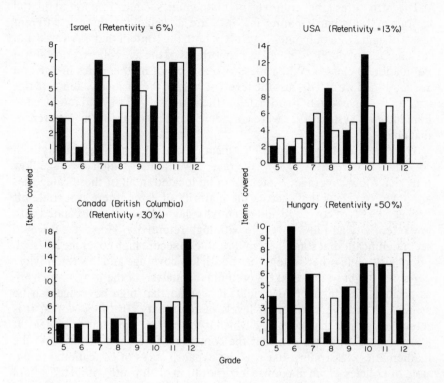

FIG 6.7.4   Low-coverage Systems: The Pace of Curricula

curriculum. Also the contrast of both of these systems with Canada (British Columbia) is very clear: in that system with its much slower-paced curriculum, the pressures on students and teachers become very different—and one outcome would seem to be high levels of retentivity, and high levels of yield in the terminal course.

We need to emphasize that the arguments and data being offered here are tentative. The responses from systems to the RQ reflected the *intended curriculum* of each system and, as has been seen in other chapters in this volume, data on an intended curriculum do not always reflect the implemented curriculum. But while this is a possibility, one that fore-shadows further work, it does seem from what we have seen here that the line of inquiry opened up by the Cockcroft Committee based on specu-lations about one system does hold across systems—and in doing so raises an important set of curricular questions for many, if not all, systems of education.

6.8 Summary

The structures and contexts of mathematics education found in the various systems participating in SIMS are associated with very different levels of aggregate outcomes in terms of both the content that is covered and the numbers of pupils, defined as proportions of the age cohort, enrolling in Population B classes. When these two aspects of mathematics instruction are combined we can discuss the levels of yield of the various systems. In this chapter we have sought to both define the various levels of yield found in the Population B systems that were considered and account for the differences that emerge.

The chapter offers a crude classification of yield; high retentivity-high coverage, high retentivity-low coverage, low retentivity-high coverage, low retentivity-low coverage. Systems can be located in all of these categories although it is clear that, across the majority of cases, there is a trade-off between levels of coverage and retentivity; high coverage is associated with low retentivity and low coverage with high retentivity. However, it would seem significant that some systems are able to secure high coverage *and* high retentivity while other systems have both low coverage and low retentivity.

As we sought to undertake a preliminary analysis of the forces which are associated with high or low yield, it seemed that high coverage can be associated with very different levels of retentivity. In other words, it is retentivity rather than coverage which is the major determinant of yield. It also seems that, as we consider the contextual forces which surround the mathematics classroom, that it is the expansiveness of educational provision, or policies which maximize incorporation of all youth within the school system, which have a primary influence on retentivity. Where more students are enrolled in school, and in particular where more girls are enrolled, the higher the retentivity of the mathematics classroom.

Such forces are also associated with the mathematics curriculum itself— and the curriculum can be seen as perhaps a major carrier of larger social forces which determine the character of the Population B classroom. High retentivity is associated with curricular structures which seem to be associated with slower *pace* in the lower secondary grades. Following the suggestion of the Cockcroft report on the teaching of mathematics in England, *Mathematics Counts*, we might suggest that such structures are often the outcome of a lack of alignment between the policies and aspirations of the educational system at large and the more slowly changing structures of the curriculum. And, interestingly, it seems that in many cases it is the *curriculum* rather than the larger policies and ideologies which determine—and perhaps carry—the shape of both the school at large and the mathematics program.

7
Summary and Implications

7.1 The Context of the Curriculum

This volume has two major goals. First, it is written to stand in its own right as a study of school mathematics curricula from an international perspective. We have described the content of the curriculum and the contextual factors that surround the curriculum. Second, we have sought to provide a backdrop for reading and interpreting the achievement data that will be reported in Volume 2 of the Study. A major finding of this volume is that while there is a common body of mathematics that comprises a significant part of the school curriculum for the two SIMS target populations (13-year-olds and those in the last year of secondary school who have continued their study of mathematics throughout schooling), there is substantial variation from system to system in the mathematical content of the curriculum. Such variation *must be taken into account as cross-cultural patterns of student achievement are examined* (Peaker 1969).

The study of the mathematics curriculum was based on the model introduced in Chapter 1. The SIMS view of the curriculum entails an examination of curricular content from three perspectives: intended, implemented and attained. Such an analysis provides a rich context for studying student outcomes (both attitudes and achievement) for it enables one to view patterns of achievement across systems, say, in the light of the content of the official or stated curriculum in each country as well as what subject matter was reported to be taught by the teachers. In other words, the SIMS model permits a "triangulation" on student outcomes from two base points: the intended and the implemented curriculum (See Figure 7.1.1). Furthermore, as the figure also indicates, for those systems that participated in the longitudinal, or classroom process, version of SIMS, two additional sources of information provide an even richer context for analyzing student outcomes: pretest data (achievement and attitudes of the students in the sampled class at the beginning of the school year) and classroom processes data (detailed information on how the subject matter was handled by the teacher when it was presented during the school year). Volume 2 in this series of international reports deals with the description and analysis of student outcome data while Volume 3 presents analyses of the concomitants of growth in mathematical achievement that are possible within the longitudinal version of SIMS.

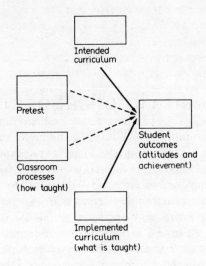

FIG 7.1.1   The SIMS Model: "Triangulation" on Student Outcomes

In Chapter 2, the role of the grid in the Study was discussed. The SIMS content-by-behavior grid is central to the Study, for it represents an international consensus on the content of the mathematics curriculum for each population. While the grid is not to be viewed as an "ideal international mathematics curriculum," it does serve to specify a sort of "international menu" of school mathematics. Even though every attempt was made to provide a comprehensive listing of topics in the grid, there are aspects of mathematics found in some systems' curricula that have not been included; for example, Infinite Sets is not in the Population A grid, yet is a significant topic at this level for some systems (e.g., Ireland). For Population B, logic was rated as important by several systems. However, for those systems for which there were data at the time of developing the grid, *this topic* did not have sufficient *international* importance to justify inclusion. In spite of such limitations, the grid does provide an operational definition of "mathematics" for the purposes of SIMS. Hence, for example, when outcomes in "Algebra" are discussed, what is meant is Algebra as defined by the international grid.

Chapter 3 describes the two target populations, Population A and Population B, and how they were embodied in each system that took part in the Study. The distinctive characteristics of each population are discussed. By and large, Population A is relatively uniform in scope from system to system in that it encompassed virtually all of the age cohort at the targeted grade level. An important difference was introduced in Japan and Hong Kong, however, in that Population A was chosen to be the grade in which the modal age was 12 years, rather than 13 years (as in the international definition). Variation was also introduced in that Population A is viewed differently with respect to placement in various systems. In a few systems, (e.g., England and Wales and The Netherlands) Population A is clearly in the second level of schooling (postprimary). In others (e.g., USA, Canada (Ontario)) Population A is regarded as being either a part of first level education, or on the "borderline" between first and second level education.

For Population B, there is much more contextual variation from system to system. In some systems (e.g., Hong Kong and Scotland), Population B spans two grades. The number of courses available to (and in many cases taken by) Population B students varies. The proportion of the age cohort remaining in secondary school of the school system is also a major factor influencing the character of Population B. The overall retentivity in secondary school ranges from a high of 92 percent (Japan) to a low of 17 percent (New Zealand). From the point of view of SIMS, however, it is the *proportion of the age group that is retained in mathematics* that is of greatest interest. These figures vary greatly from system to system as well. In Hungary, *all* pupils who remain in school enroll in Population B mathematics. Since 50 percent of Hungarian youth remain through secondary school, that system has a 50 percent enrollment rate in Population B mathematics—the highest by far of any SIMS country. In England and Wales and Israel, it was estimated that only 6 percent of the age cohort were in Population B mathematics. The median retention rate in Population B mathematics for all of the SIMS countries was about 12 percent of the age cohort.

With respect to class size, considerable variation was also found between systems at both target population levels. For Population A, the median number of students ranged from lows of 19 in Luxembourg, 20 in Belgium (French) and 21 in Belgium (Flemish) to highs of over 40 students in Hong Kong, Japan, Swaziland and Thailand. In Population B, class size was typically smaller than for Population A. England and Wales reported average enrollments of 10 students and Belgium (French) 13 students while Japan and Thailand maintain large classes at this level with median class size of 43 and 40, respectively.

7.2 The Content of the Curriculum: Population A

The content of the intended curriculum is dealt with in Chapter 4 and of the implemented curriculum in Chapter 5.

7.2.1 Commonality in the Population A Curriculum

The greater part of the content of the international item pool for Measurement and Arithmetic is included in the *intended* curriculum for all systems. An exception is for Measurement; France, Flemish and French

Belgium and Luxembourg found only about 70 percent of this content to be appropriate for their curricula. In the remaining systems, the appropriateness rating was 80 percent or more. For Arithmetic, all systems reported 80 percent or more of the content to be appropriate.

Algebra was generally judged to be a part of each system's Population A curriculum. But there are some topics within Algebra (for example, Polynomials and Rational Expressions, Finite Sets and Integer Exponents) that were in the curricula of only from one-third to one-half of the systems.

At the level of the *implemented* curriculum, we also see a common core of mathematical content that is taught to the majority of students in each system. By and large, this core corresponds to the mathematics that is intended to be taught (that is, Arithmetic and Measurement). *Algebra* is less of a common core area, but even for this topic, the majority of the systems had coverage indices of more than 70 percent. The topics of Integers and Rational Numbers were especially well covered (on average, teacher coverage was more than 85 percent).

Two clusters of systems were identified based on the emphasis placed on Algebra in the intended curriculum for Population A. The majority of systems were in the "High Algebra" cluster. For those in the "Low Algebra" cluster, explanations were sought in the contextual data for those systems. In Sweden, two explanations were offered. First, it was noted that in Sweden Population A is only the seventh year in school. Therefore, one might conjecture, the curriculum is still influenced by an orientation toward primary school arithmetic. As a counter example, however, the case of Japan might be offered. In that country the Population A curriculum provides an intense introduction to Algebra even though the students are only in their seventh year of schooling. It was also noted in Sweden that two mathematics programs are available to Population A students—a long and a short course. Algebra is much more a part of the Swedish long course than the short course (see Chapter 5, Section 5.4).

In the United States at Population A there is a multiplicity of courses, with only the course for the most able students providing a substantial amount of Algebra. This differentiation between classes contrasts greatly with Japan, as Figure 7.2.1 indicates. The box-plots show, first of all, somewhat comparable implemented coverage for Arithmetic in the two systems and lower median coverage of Algebra in the two systems. But a dramatic difference between the systems is found in the between-class-variation in implemented coverage of Algebra.

Figure 7.2.1 indicates, for example, that classes in the United States ranged in coverage from *none* of the Algebra items taught to *all* of them taught. Japanese classes, on the other hand, had Algebra coverage that was on average high and differed little from class to class. That is to say, Population A students in the United States have quite different opportunities to learn Algebra, while those in Japan have relatively equal access to the subject.

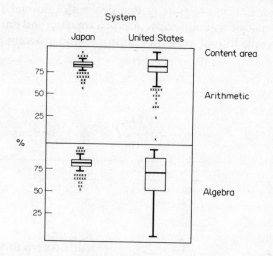

Fig 7.2.1   Opportunity to Learn Arithmetic and Algebra in the United States (Eighth Grade) and Japan (Seventh Grade)

7.2.2   Curricular Diversity—Population A
Geometry
Intended Curriculum

The SIMS systems show very little agreement as to what topics in Geometry should be in the Population A curriculum. The stem-and-leaf table below illustrates the high between-system variation in geometric content in the intended curriculum. But even in the face of this great

Table 7.2.1   *Population A: Stem-and-Leaf Table for Intended Coverage of Geometry*

Population A	
Stem	Leaf
9	NZE JPN SCO
8	ENW HUN CON
7	FIN NGE ISR THA NTH
6	SWA HKO
5	SWE USA
4	CBC FRA
3	IRE
2	BFL BFR LUX

diversity, there is, with the exception of Luxembourg and the two Belgian systems, a small core of topics common to all of the SIMS countries: Classification of Plane Figures, Properties of Plane Figures, Congruence of

Plane Figures, Coordinates and some work on Simple Deduction. Beyond that, there is a cluster of systems that deal with Informal Transformations, Spatial Visualization and Solids. A much smaller (and different) cluster of systems provide formal work on Transformational Geometry.

Implemented Curriculum

Table 7.2.2   *Population A: Stem-and-Leaf Table for Implemented Coverage of Geometry*

Stem	Leaf
10	
9	
8	SWA HUN
7	
6	NZE NGE NTH
5	CBC CON JPN ENW THA
4	ISR FRA USA CON
3	BFL LUX SWE FIN
2	
1	
0	

Implemented coverage for Geometry is on average for all countries only 50 percent, ranging from a high of 87 percent for Hungary to a low of 31 percent for Belgium (Flemish). It is interesting to note that Hungary's high implemented coverage corresponds to their intended coverage index of 88 percent, as does Belgium's low coverage index. Japan's implemented coverage index of 51 percent, however, is low in light of its intended coverage index of 90 percent.

The great diversity in Geometry is suggested by the patterns of teacher coverage for the item shown in Figure 7.2.2. The content of this item, offering a rather formal approach to vectors (especially for Population A) was taught to 78 percent of the classes in Belgium (Flemish) and to few classes in any other system (for example, 17 percent in France, and 10 percent or less in all other systems).

The Bourbaki program, as the above item suggests, provides an alternative to more conventional content as represented by a Euclidean approach to the subject. Little evidence is provided by SIMS, however, that this novel, distinctive view of mathematics has been widespread across systems either in terms of its presence in the curricula of systems outside of the small cluster identified above, or in terms of the extent to which teachers are presenting the material to their students. At the time of this Study, the issue of what kind of geometry is most appropriate for lower secondary school students was still very much an open question.

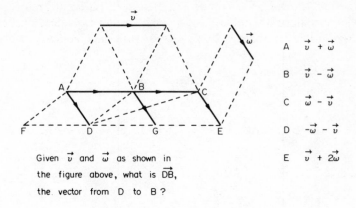

A $\vec{v} + \vec{\omega}$

B $\vec{v} - \vec{\omega}$

C $\vec{\omega} - \vec{v}$

D $-\vec{\omega} - \vec{v}$

E $\vec{v} + 2\vec{\omega}$

Given $\vec{v}$ and $\vec{\omega}$ as shown in the figure above, what is $\overrightarrow{DB}$, the vector from D to B?

FIG 7.2.2 Geometry Item (Population A)

In Statistics, most countries seem to cover topics in basic descriptive statistics, with little work done in probability in any country at the Population A level. While mathematics educators and other curriculum specialists offer a unanimous voice in urging the inclusion in the curriculum of topics in exploratory data analysis and elementary probabilistic concepts, little evidence was found in any system of widespread implementation of such topics in schools at this level.

7.2.3 Correspondence between Intended and Implemented Curriculum: Population A

An indication of the extent of correspondence between the content of the intended curriculum and the implemented curriculum may be gleaned from Figure 7.2.3. The data points, representing systems, show the indices of intended and implemented coverage. The figure is divided into four quadrants, using the means of each index as a rough dividing line between the systems. Therefore, for example, systems in Region I are those at or above the means on both intended and implemented coverage.

Region I: High Intended – High Implemented
Systems: Japan, Hungary, France and Canada (British Columbia).
These systems had high intended coverage of Algebra, and high opportunity-to-learn this subject matter.

Region II: High Intended – Low Implemented
Systems: Belgium (Flemish), The Netherlands, New Zealand and England and Wales.

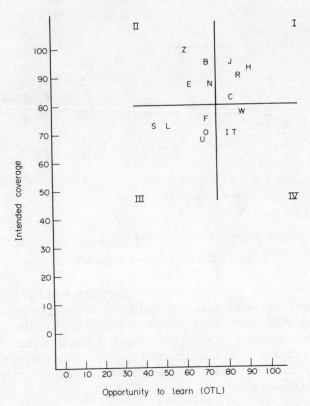

FIG 7.2.3   Intended Coverage vs. OTL for Algebra: Population A

For these systems, the National Centers reported an above-average proportion of the Algebra items as appropriate for Population A. However, teacher coverage of the content of these items was below the international average.

Region III: Low Intended – Low Implemented
Systems: Finland, Luxembourg, Sweden, Canada (Ontario) and United States.
In these systems, teacher coverage of the Algebra items was low—in accordance with the below-average ratings for appropriateness that were provided by the National Centers. Apparently, for this cluster of systems, Algebra (or, at least, the Algebra of the SIMS pool) is not a topic of considerable importance for Population A.

Region IV: Low Intended – High Implemented
Systems: Swaziland, Israel and Thailand.
In this final cluster of systems, National Center ratings suggest that

Algebra is judged as not appropriate for Population A. The teachers, on the other hand, reported relatively high coverage of this subject matter. The reasons for this lack of congruence between National Center expectations and teacher coverage warrant further exploration. One reason may be that the National Center ratings are underestimates of the content of the curriculum. In the case of Israel, for example, none of the Statistics items was rated as being appropriate for the Population A curriculum; the opportunity-to-learn data, however, indicated coverage of over 50 percent, with items dealing with Graphing and Interpreting Data nearing a coverage of 60 percent. (See Figure 7.2.4.)

FIG 7.2.4   Item by Class Display for Statistics: Israel (Population A)

Region II of Figure 7.2.3 is worthy of note since teacher coverage is low compared to the scope of the intended curricula for those systems. Factors accounting for the lower teacher coverage would include a consideration of the amount of time allocated to the subject, as well as a look at the existence of multiple courses, or tracking.

In summary, with respect to Population A, we have analyzed the content of the curriculum both in terms of what is intended and what is taught and have found considerable variation between systems in the mathematical "meal" that is made available to students. We have, on the one hand, some systems that offer a rather rich fare, characterized by relative intense coverage of topics in Algebra, Statistics, and to some extent in Geometry, as well. For other systems, a less "robust" fare is available. There is, to be sure, extensive coverage of Arithmetic and Measurement but these topics are, in many cases, "more of the same" material that has been presented in the primary grades.

7.3   The Content of the Curriculum: Population B

7.3.1   Commonality in the Population B Curriculum

There is a substantial common core of mathematics at Population B. Since at the Population B level there is in most systems an emphasis on preparing students for further work in mathematics and allied fields in institutions of advanced study, this common purpose seems to have brought about a commonality of curriculum which is most dramatically seen in the content areas of Algebra and Elementary Functions and Calculus. Algebra has high intended coverage on average (over .90) for all systems, while Elementary Functions and Calculus is high except for Thailand (.60) and Canada (British Columbia) (.30).

In terms of opportunity-to-learn, Algebra has an overall mean index of about .85, with all systems above .80 except Israel (.56) and Thailand (.70). Similarly, coverage of Functions and Calculus is high except in Israel, Thailand, the United States and Canada (British Columbia) where teacher coverage is .70 or below.

In the plane E, T_v is the translation corresponding to the vector v, S_Ω is the half-turn about the centre of Ω, $h(\Omega,\kappa)$ the size transformation of magnitude κ and center Ω. If

$$f = h(\Omega,3) \circ T_{\vec{r}} \circ T_{\vec{r}+2\vec{r}} \circ h(\Omega,-\tfrac{1}{3}) \circ S_\Omega$$

then f is

    A   a half-turn about Ω
    B   a half-turn about a point other than Ω
    C   a translation
    D   a size transformation of magnitude -1
    E   a reflection in a line through Ω

Figure 7.3.1   *Geometry item (Population B)*

7.3.2   *Curricular Diversity—Population B*

Geometry exhibits considerable diversity in Population B, as it did in Population A. While there is some commonality among the systems with respect to Euclidean Geometry and Trigonometry, there is great variation in the implemented coverage of Vector Methods (ranging from .07 for Israel to .81 for Belgium (Flemish)) and Transformational Geometry (ranging from .04 for Israel to .63 for Belgium (Flemish)).

The item in Figure 7.3.1 is illustrative of the coverage accorded Transformational Geometry at Population B. Teacher data indicated that 88 percent of the classes in Belgium (Flemish) and 81 percent of those in Japan were taught the content of this item. By contrast, virtually no classes in Israel, Sweden, Canada (British Columbia) or the United States were taught this subject matter.

For Finite Mathematics (Combinatorics), implemented coverage ranges from .10 for Canada (British Columbia) to .99 for Japan. It should be kept in

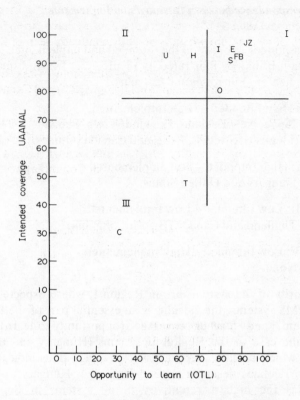

FIG 7.3.2   Intended Coverage vs. OTL for Elementary Functions/Calculus:
Population B

mind that for this content area there were only 4 items in the SIMS pool.

For Probability and Statistics, indices range from .17 for Israel to the .80s for Finland, Japan, New Zealand and Thailand.

The element of diversity in Population B mathematics provides potentially important data for those concerned with curricular experimentation and innovation. The topics of Probability and Statistics, while available in some systems (for example, New Zealand and Finland) are less available (implemented coverage is lower) in Canada (British Columbia) and Hungary. On the other hand, the former systems are rather selective in terms of their enrollments in Population B mathematics while the latter systems retain relatively large proportions of students in mathematics. It would seem to be important to explore the extent to which topics in Probability and Statistics can be incorporated in the curricula of "more retentive" countries—and vitally important to consider, for example, the role that technology, and in particular computer technology, could play in the effective teaching of these topics.

7.3.3   Correspondence between Intended and Implemented Curriculum: Population B

The correspondence between the intended and implemented curriculum for Population B may be illustrated by Figure 7.3.1 which displays system data for Elementary Functions and Calculus.

Region I: High Intended – High Implemented
Systems: Japan, New Zealand, England/Wales, Finland, Belgium (Flemish), Sweden, Israel and Canada (Ontario).

Region II: High Intended – Low Implemented
Systems: Hungary and United States.

Region III: Low Intended – Low Implemented
Systems: Thailand and Canada (British Columbia).

Region IV: Low Intended – High Implemented
Systems: None.

The majority of the systems are in Region I, where expected. That is, for most SIMS systems, the calculus is an essential part of the Population B curriculum, and teacher coverage (opportunity-to-learn) is high. However, the case of the Region II systems (Hungary and the United States), in which teachers report lower coverage than indicated in the intended curriculum, deserves further discussion. Hungary, it should be recalled, has the highest retentivity of all systems in Population B mathematics. All students who are in the final year of secondary school in Hungary are enrolled in Population B mathematics. In the case of the United States, not all Population B classes are taught calculus and, as was noted earlier, only about 20 percent of the classes take a "full-fledged" course in calculus (Crosswhite, et al. 1985).

In the Region III systems little calculus is in the Population B curriculum for either Thailand or Canada (British Columbia). Correspondingly low teacher coverage is reported.

7.4   Yield

One of the most provocative findings reported in the previous chapters centers on the notion of the *yield* of school mathematics education. As already noted, the proportion of the age cohort enrolled in Population B ranges from a high of 50 percent in the case of Hungary to a low of 6 percent in the cases of England and Wales and Israel. The levels of coverage of the total body of mathematics specified in the SIMS item pool ranges from 51 percent in the case of Canada (British Columbia) to 91 percent in the case of Japan. When we combine these separate notions of coverage and retentivity and see them as different components of yield, we see considerable variation in the outcomes of teaching in the various systems. In almost all systems *all* students study mathematics in the early years of secondary school. However, as students progress through the secondary school, the systems differ greatly in terms of the proportions that continue to enroll in mathematics courses—in particular, with respect to SIMS, those students enrolled in Population B courses. In Chapter 6 it was suggested that these outcomes can be classified in terms of a two-way table using teacher coverage of the total item pool as a criterion. We see Belgium (Flemish), England and Wales, Japan, New Zealand and Sweden falling into a high coverage-low retentivity cell (retentivity less than 13 percent of the age cohort), and Canada (British Columbia), Hungary and Canada (Ontario) falling into a low coverage-high retentivity cell and Finland (with a retentivity index of 15 percent and the coverage index of 85 percent) falling into the high coverage-high retentivity cell. What lies behind such differences in yield?

It was also suggested in Chapter 6 that perhaps the principal factor contributing to a system's coverage index was the level of its coverage of Elementary Functions and Calculus. Almost all systems (the exceptions are Hungary, Israel and Canada (Ontario)) expect to teach most of the Algebra-related material in the SIMS pool. The difference between systems is found in the coverage of Elementary Functions and Calculus where there is substantial variation in the levels of coverage with Canada (British Columbia), the US and Hungary covering less than 70 percent of the pool and the remaining systems covering more than 80 percent of the content in this area. And among the school systems in industrialized nations, coverage of Elementary Functions and Calculus is associated with the number of students included within the Population B cohort. For example, Canada (British Columbia) enrolls 30 percent of the age cohort in Population B and has low coverage of Elementary Functions and Calculus.

However, there are systems that enroll comparatively larger proportions of their young people in Population B mathematics and do cover Elementary Functions and Calculus with the same level of intensity as do those who enroll many fewer students. Thus, England and Wales, which enrolls 6 percent of its age cohort in Population B mathematics, and Finland, which enrolls 15 percent of its students in Population B, both have coverage indices for Elementary Functions and Calculus of .88, Canada (Ontario) with 19 percent of it students enrolled in Population B has a coverage index of .83. How can we account for these differences?

As was suggested in Chapter 6, part of the explanation of such different levels of enrollment in mathematics classes that can be termed high coverage is associated with the strength of what might be termed a tradition of selectivity within the different systems. Such *selectivity* is associated with lower levels of enrollment within both the upper secondary and the tertiary systems. Furthermore, one would expect close linkages between the kind of upper secondary schooling that has mathematics close to its core and a biasing in favor of mathematics from schools that draw enrollments from predominantly professional/managerial households.

Such factors are however external to the school system and it is not entirely clear how these forces work themselves out within the school system. Particularly intriguing is the persistence of such selectivity and social bias which is, of course, closely associated with the traditions of selective secondary education which marked the first half of this century. But almost all of the school systems involved in SIMS have had policies in place designed to open access to the school for 20 or more years.

7.4.1   Illustrative Yield Data

The content areas of Algebra and Elementary Functions and Calculus are taken as representatives of Population B mathematics. Table 7.4.1 gives

Table 7.4.1   *Intended and Implemented Yield for Population B Mathematics*

	Algebra		
System	Pop B Retentivity	Intended Coverage	Implemented Coverage
Belgium (Flemish)	.10	1.00	.92
Canada (British Columbia)	.30	.84	.83
Canada (Ontario)	.19	1.00	.92
England/Wales	.06	1.00	.87
Finland	.15	.96	.92
Hungary	.50	.92	.86
Israel	.06	.96	.72
Japan	.12	.92	.99
New Zealand	.11	1.00	.92
Sweden	.12	.92	.90
United States	.13	.88	.89

	Elementary Functions and Calculus		
System	Pop B Retentivity	Intended Coverage	Implemented Coverage
Belgium (Flemish)	.10	.96	.89
Canada (British Columbia)	.30	.39	.32
Canada (Ontario)	.19	.83	.83
England and Wales	.06	.98	.88
Finland	.15	.96	.88
Hungary	.50	.96	.68
Israel	.06	.98	.79
Japan	.12	.98	.92
New Zealand	.11	1.00	.94
Sweden	.12	.89	.86
United States	.13	.93	.57

coverage indices in these content areas for systems for which complete data were available.

This coverage and retentivity data for selected systems is graphed in Figures 7.4.1 and 7.4.2 for Algebra and Elementary Functions and Calculus.

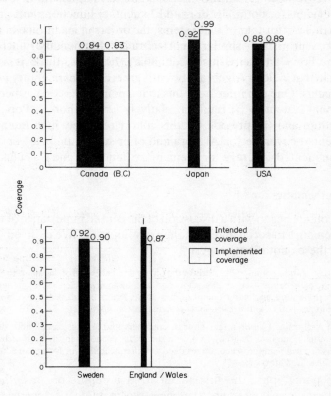

FIG 7.4.1 Intended and Implemented Coverage for Algebra for Five Countries

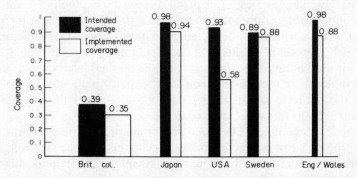

FIG 7.4.2 Intended and Implemented Coverage for Analysis (Elementary Functions and Calculus) for Five Countries

The widths of the histogram reflect Population B retentivity. We see, for example, that Canada (British Columbia) has high retentivity and England and Wales low retentivity. In accordance with intuition, coverage appears to be negatively related to retentivity. That is, higher retentivity is associated with low coverage at both the intended and the implemented levels. This trend is evident for both Algebra and Elementary Functions and Calculus.

Nevertheless, for four of the systems, the orders of magnitude of the yield indices are surprisingly similar for Algebra and Elementary Functions and Calculus. For Canada (British Columbia), however, there is a striking contrast in that yield in Algebra is twice as great as for Elementary Functions and Calculus. One can infer from this that a policy decision has been in this educational system to (i) retain a relatively large cohort in Population B mathematics and (ii) provide for this cohort relatively high intended and implemented coverage for Algebra and offer substantially lower intended and implemented coverage for Elementary Functions and Calculus.

7.5   Mathematics for All

This volume began with a discussion of the central importance of the study of mathematics to social and economic development. We close on the same note, as these quotations suggest.

Mathematics is fundamental to the study of the physical sciences and of engineering of all kinds. It is increasingly being used in medicine and the biological sciences, in geography and economics, in business and management studies. It is essential to the operations of industry and commerce in both office and workshop. (*Mathematics Counts*, 1982)

Without a well-trained labor force, enough engineers and strong science education in our school systems, none of us is going to have a successful economy in the high-technology era ahead of us. To work, modern economies need a well-educated labor force. An educated elite will not suffice. L. Thurow (1985)

These paragraphs, the first taken from a report of an authoritative national commission set up to review the mathematics curriculum, the second taken from a discussion of economic and social futures by a respected economist, capture views of mathematics being taught in the school that are increasingly being expressed in many societies. Two themes recur in the explorations of the consequences that seem to follow such starting points: first, that the mathematics being taught in most school systems has not kept pace with changes in mathematics itself and in the ways in which mathematics is used in business, industry, and the marketplace; second, that too many students are terminating their study of mathematics too soon—or are being allowed to terminate their study too soon—given the importance of the field in both later schooling and in the workplace.

Although the emphasis between one or another of these starting points varies, the theme of interaction between economic and social developments and the curriculum is one that is increasingly shared both within and without the field of mathematics education. Furthermore, such a theme leads to arguments for new forms of a "core curriculum" that gives increasing attention to the "needs" of the world of work. In such a curriculum mathematics and science have central places: they provide the foundation of the specific training of the very large number of technicians and technologists required by modern orders, they provide the basis for the general skills and understandings which will be increasingly required by all participants in "modern" workforces, and they provide the basis for the general skills and understandings that will be required for reflective participation in societies that are increasingly dominated by scientific and technological concerns. Within mathematics education these forces are bringing about a new set of imperatives for the field: more mathematics must be taught to more students—in other words a call for "mathematics for all."

"Mathematics for all" is of course a slogan and represents a concern that is not new to the field. Keitel (1987) has suggested, for example, that "mathematics for all" is the theme that has characterized "the most significant achievements of mathematics education in this century"—but, as she goes on to note, these achievements have been "alas, intellectual achievements, *not achievements in the schools.*" It is the mathematics of the *schools* and its fit with the needs of majorities and not minorities that is seen as defining what is perhaps the major agenda for mathematics education for the balance of this century (Howson and Wilson, 1986).

This emerging agenda for mathematics education must, of course, be set against the backdrop of fundamental changes in the role of mathematics in the school that have already taken place in comparatively recent years. In Europe, for example, mathematics was closely associated with the curriculum of the academic secondary (i.e., grammar-type) school until the advent of comprehensivization in the 1950s and 1960s. For the majority of students who did not attend such schools, little or no mathematics (as distinct from arithmetic) was taught. In the school systems of North America, where universal secondary education emerged as an ideal in the 1930s and 1940s, introductory algebra and sometimes geometry were offered to all students earlier than in Europe. But even in those systems assumptions about the abilities required to complete advanced mathematics meant that only a minority continued with the subject into the upper secondary school. It is this association between mathematics and an academic education oriented toward university programs in the physical sciences and engineering that has created and continues to create the pervasive problems which will have to be faced as the implications of notions like mathematics for all are worked out. Thus, Usiskin (1987), writing in the context of the United States, notes that although school authorities are mandating increased numbers of courses in mathematics as a response to concerns about increased numeracy, the one million high school graduates who take the Scholastic Aptitude Test, a widely used test taken by college- and university-bound students, already report taking more than three of the possible four courses in mathematics that are available in the American 4-year high school. The implication is that any new requirements for mathematics course-taking that might be imposed would affect only those students who are not already meeting such requirements, i.e., students who are predominantly not university-bound. Usiskin goes on to comment that "consequently students with a record of poorer academic performance are winding up in courses intended to provide preparation for college and university, courses in which they have little interest. Furthermore, high school mathematics courses are characteristically among the more severely graded courses with the highest failure rates." Damerow's and Westbury's (1985) metaphor captures the ambience of such contexts very aptly;

Many are invited to the table to begin a meal, but few are given the opportunity to finish—because much of what is served is unattractive or inedible.

This general picture of the selection processes that occur in the mathematics classroom is heightened further when such factors as gender and social class are added to the picture. The traditional mathematics classroom in the upper secondary school has been—and in some societies still is— dominated by male students. Equally significantly, mathematics has often been a subject one of whose tasks in the school has been of sorting of students into different tracks and streams having very different occupational and social implications. Adda (1986), for example, reports that in France only 16 percent of the 17-year-old cohort born in 1962 remained in the mainstream mathematics program and that the process of selection which produced this outcome overwhelmingly disadvantaged socially-deprived families. In 1976–77 52 percent of pupils from families whose fathers were "upper executives" were in the C-stream of the French *lycée* whereas only 6 percent of the children of workers were in this stream. Adda concludes that "mathematics teaching as it is practised today is not neutral but produces a correlation between school failure in mathematics and the sociocultural environment." (p. 58)

Such findings have been found in many school systems and are increasingly being seen as an outcome of a bias which inheres in the mathematics curriculum itself, and the traditions associated with that curriculum. In a paper reviewing the issues in mathematics education for the 1990s commissioned for the International Commission on Mathematics Education, Howson, Nebres and Wilson (1985) note that "the canonical mathematical curriculum was developed . . . with a minority of the society in mind—for only an elite sector had access to a substantial number of years of schooling" at the time when that curriculum was developed. The curriculum that emerged in this context was one whose standards were those thought appropriate to the "needs" of this minority and the standards that were embedded in this curriculum were the standards that were attainable by that minority. Thus, writing in 1945, when 39 percent of the age cohort was graduated from the American high school (*Statistical History*, 1976), the Harvard University Committee on General Education in a Free Society claimed in their discussion of the school program that:

We must recognize that for the mathematically less gifted pupils in the ninth grade there is little straightforward mathematics available beyond elementary instruction in arithmetic and informal geometry which . . . should include guidance in the use of formulae, equations, graphs, and right angle trigonometry . . . On the other hand it is of course desirable to stimulate the interest of mathematically inept students in the number relations of arithmetic and the elementary principles of geometry by presenting mathematics in various disguises—such as shop mathematics, business arithmetic, mathematics of the farm, and so on. In such novel forms these students can be brought to reexamine and improve their grasp of simple arithmetic and its application in practical problems. (Harvard University 1985: p. 163.)

The report had previously noted that "probably little more than half the pupils enrolled in the eighth grade can derive genuine profit from substantial instruction in algebra or can be expected to master demonstrative geometry." (p. 161)

It is only in recent years that the assumptions that legitimated such views have been opened to widespread questioning and the linkages between the curriculum and the seeming reality of the "empirical" facts that underlie the kind of judgments offered by the Harvard Report. Thus, as we have noted above, *Mathematics Counts* (1982) pointed out that, despite the changes in the structure of schooling in England and Wales that have made mathematics a subject for all, the "deep structure" of the English mathematics curriculum did not change, and has not changed for 40 or more years. The results have been a situation in which the implicit pace of instruction in mathematics is too fast for many of the "new" pupils in the "new" secondary school. In other words, it is the curriculum and the forces which circle around the curriculum rather than a school organization as such that stand between overall policies (which may, for example, emphasize opportunity, or goals such as mathematics for all) and the effects of these policies.

7.6 Postscript

The Second International Mathematics Study, as other IEA studies, has had the unique opportunity of capturing the "natural variation" that occurs across school systems and the cultures in which these systems are embedded. There are varieties in the contexts of education. We have found varying degrees of school retentivity, of between-and-within school differentiation of curricula, and of size of classes. With respect to the content of the curriculum, we have identified a substantial common core of mathematics that is both intended to be taught and is reported to be taught, thus affirming the existence of the "canonical curriculum" alluded to by Howson, Nebres and Wilson (1985).

There are also important between-system differences in curricular content, and from those differences we have learned some lessons about curricular implementation. We have seen the pocket of curricular innovation in those European countries in the Bourbaki tradition. While that reform movement impacted a few systems in a dramatic way, the influence was limited to their boundaries. It will be important to study the nature of that movement—its content and its sociology—in order to identify reasons for its limited influence.

As systems plan for new directions in mathematics education, data from SIMS may provide bases for case studies in curriculum design and implementation. The case of Hungary is important in that at the Population B level, all those who remain in school are enrolled in advanced mathematics. Consequently, Hungary has the highest retentivity of any SIMS system. The case of Canada (British Columbia), however, provides another important set of data. This system is second highest in retentivity in Population B mathematics (30 percent, as compared with 50 percent for Hungary), but has identified only a subset of the topics in advanced mathematics for this large cohort of students—with Calculus *not* a part of the advanced mathematics program. However, for advanced Algebra, the extent of coverage given to this topic, coupled with the high retentivity for Population B, results in a high yield in Algebra for Canada (British Columbia).

The variations displayed by the SIMS data raise intriguing scenarios for cross-national curricular experimentation. In particular, it might be asked whether there are topics in advanced mathematics (such as finite mathematics, for example) that are emerging in importance in this technological age that might usefully be made available to increasing numbers of students in the "low yield" systems—topics that would serve to update the mathematics that is provided in schools. Furthermore as Howson and Wilson (1986) have noted, "Never in the history of mathematics education has any development opened up such a vast range of possibilities and challenges to the educator as has the microcomputer" (p. 68). For example, since computers are essentially discrete machines, there has been in recent years increased interest in the study of discrete mathematics at both post-secondary and secondary school levels. It is interesting to note that in the late 1970s, as the SIMS grid was being developed, it was the consensus that Finite Mathematics as a topic had so little international importance that only a few items on combinatorics were included in the SIMS pool. Presently, however, there is a great deal of interest in the topic and in some countries there are pressures to redirect attention from calculus to the study of discrete mathematics.

As efforts are made to make school mathematics available to more and more students, it is critically important that the content provided is matched to the educational needs of those taught. It is reasonable to expect, even to insist, that school mathematics curricula be continually assessed and analyzed to help ensure their integrity and suitability for current and future educational purposes.

Today we maintain ourselves. Tomorrow science will have moved forward one more step, and there will be no appeal from the judgment which will then be pronounced on the educated. A. N. Whitehead (1929)

References

Adda J 1986 Fight against academic failure in mathematics. In *Mathematics for All: Problems of Cultural Selectivity and Unequal Distribution of Mathematical Education and Future Perspectives on Mathematics Teaching for the Majority*. Science and Technology Education Document Series No. 20, Division of Science, Technical and Environmental Education, UNESCO, Paris, pp. 58–61

Ball S J 1981 *Beachside Comprehensive: A Case Study of Secondary Schooling*. Cambridge University Press, Cambridge

Bloom B S 1974 Implications of the IEA-studies for curriculum and instruction. *School Review*. 82:413–435

Bourdieu P, J-C Passeron 1977 *Reproduction in Education, Society and Culture*. (Trans. R Nice). Sage Publications, London and Beverly Hills, CA

Burstein L (Ed) 1989 *The IEA Study of Mathematics III: Student Growth and Classroom Process in Lower Secondary Schools*. Pergamon Press, Oxford

Clark B 1985 *The School and University: An International Perspective*. University of California Press, Berkeley, California

Cliff N 1983 Evaluating Guttman scales: Some old and new thoughts. In: Wainer H, S Messick (eds) *Principles of Modern Psychological Measurement*. Lawrence Erlbaum, Hillsdale, NJ, pp. 283–299

Comber L C, J P Keeves 1973 *Science Education in Nineteen Countries*. International Studies in Evaluation, Vol. 1. Almqvist and Wiksell, Stockholm

Damerow P, I Westbury 1986 Conclusions drawn from the experiences of the New Mathematics movement. In *Mathematics for All: Problems of Cultural Selectivity and Unequal Distribution of Mathematical Education and Future Perspectives on Mathematics Teaching for the Majority*. Science and Technology Document Series No. 20, Division of Science, Technical and Environmental Education, UNESCO, Paris, pp. 22–25

Damerow P, M E Dunkley, B F Bienvenido, B Werry (eds) 1986 *Mathematics for All: Problems of Cultural Selectivity and Unequal Distribution of Mathematical Education and Future Perspectives on Mathematics Teaching of the Majority*. Science and Technology Education Document Series No. 20, Division of Science, Technical and Environmental Education, UNESCO, Paris

Davis R B 1984 *Learning Mathematics: The Cognitive Science Approach to Mathematics Education*. Croom Helm and Ablex, London and Norwood, New Jersey

Evans J 1985 *Teaching in Transition: The Challenge of Mixed Ability Grouping*. Open University Press, Milton Keynes

Fensham, P J 1980 Constraint and autonomy in Australian secondary science education. *Curriculum Studies*. 12(3): 189–206

Freudenthal H 1975 Pupils' achievement internationally compared—the IEA. *Educational Studies in Mathematics*. 6: 127–186

Garden R A 1987 *Second IEA Mathematics Study: Sampling Report*. Center for Statistics, United States Department of Education, Washington DC

Harnisch D L 1984 Females and mathematics: A cross-national perspective. In *Advances in Motivation and Achievement* Vol. 2. JAI Press, Greenwich, CT

Harnisch D L, R L Linn 1981 Analysis of item response patterns: Questionable test data and dissimilar curriculum practice. *Journal of Educational Measurement*. 18(3): 133–46

Harnisch D L, M W Steinkamp, S–L Tsai, H J Walberg 1986 Cross-national differences in mathematics attitude and achievement among seventeen-year-olds. *Int. J. Educational Development*, 6: 23344

Harnqvist K 1975 The International Study of Educational Achievement, *Review of Research in Education 3*. F. E. Peacock Publishers, Inc., Itasca, IL., pp. 85–109

224

Harvard University, Committee on the Objectives of Education in a Free Society (1947) *General Education in a Free Society: Report of the Harvard Committee.* Harvard University Press, Cambridge, MA

Hirstein J 1980 From Rouaymont to Bielefeld: A twenty-year cross-national survey of the content of school mathematics. In: Steiner H G (ed) 1980 *Comparative Studies of Mathematics Curricula: Change and Stability 1960–1980.* Materialen und Studien Band 19. Institut für Didaktik der Mathematik der Universität Bielefeld, Bielefeld, FRG, pp. 55–89

Holmes B 1983 *International Handbook of Education Systems* Vol. 1: Europe and Canada. John Wiley and Sons, New York

Howson A G, B F Nebres, B Wilson 1985 *School Mathematics in the 1990s.* Paper prepared for ICMI Symposium on School Mathematics in the 1990s, Kuwait Center for Mathematics Education. School of Mathematical Studies, University of Southampton, Southampton

Howson A G, B Wilson 1986 *School Mathematics in the 1990s.* Cambridge University Press, Cambridge, England

Husén T 1967 *International Study of Achievement in Mathematics*, Vols. 1 and 2. Almqvist and Wiksell, Stockholm and John Wiley, New York

Husén T, T N Postlethwaite 1985 *International Encyclopedia of Education: Research and Studies* Vol. 8. Pergamon, Oxford

Inkeles A 1977 The International Evaluation of Educational Achievement: A review of *International Studies in Evaluation* (9 volumes) by the International Association for the Evaluation of Educational Achievement. *Proceedings of National Academy of Education* 4: 139–200

Keitel C 1987 What are the goals of mathematics for all? *Curriculum Studies* 19: 393–408

Kifer E, R Wolfe, W Schmidt 1985 Cognitive growth in eight countries: Lower secondary school. University of Illinois at Urbana-Champaign, Urbana, Illinois, Mimeo

Livingstone I D 1986 *Perceptions of the Intended and Implemented Mathematics Curriculum.* Report on the Second International Mathematics Study. Center for Statistics, United States Department of Education, Washington DC, *Mathematics Counts.* (1982) Report of the Committee of Inquiry into the Teaching of Mathematics in Schools. Her Majesty's Stationery Office, London

McKnight C et al. 1987 *The Underachieving Curriculum: Assessing U.S. School Mathematics from an International Perspective* Stipes Publishing Company, Champaign, Illinois

National Council of Teachers of Mathematics 1971 International study of achievement in mathematics. *Journal Res. Math.* 2: 69–171

Organization for European Economic Cooperation, Office for Scientific and Technical Personnel 1961 *New Thinking in School Mathematics.* Organization for European Economic Cooperation, Paris

Peaker G 1969 The international study of achievement in mathematics: A few examples of points of special interest to teachers. Int. Rev. Ed. 15(2): 222–228

Peaker G F 1975 *An Empirical Study of Education in Twenty-One Systems: A Technical Report.* Vol. 8. Almqvist and Wiksell, Stockholm and John Wiley, New York

Postlethwaite T N 1967 *School Organization and Student Achievement: A Study Based on Achievement in Mathematics in Twelve Countries.* Stockholm Studies in Educational Psychology 15. Almqvist and Wiksell, Stockholm

Postlethwaite T N 1971 Item scores as feedback to curriculum planners: A simple case from the Swedish comprehensive school and a more general model. *Scand. J. Educ. Res.* 15: 123–136

Purves A C, D L Levine 1975 *Educational Policy and International Assessment: Implications of the IEA Surveys of Achievement.* McCutchan, Berkeley, California

Purves A C 1987 The evolution of the IEA: A memoir, *Comp. Ed. Review* 31(1): 10–28

Ralston A, G Young 1983 *The Future of College Mathematics.* Springer-Verlag, New York, Heidelberg

Robitaille D F, R A Garden 1988 *The IEA Study of Mathematics II: Contexts and Outcomes of School Mathematics.* Pergamon Press, Oxford

Schildkamp-Kündiger E 1982 *An International Review of Gender and Mathematics.* Mathematics Education Information Report. ERIC Science, Mathematics and Environmental Education Clearing House, Ohio State University, Columbus, Ohio

Schildkamp-Kündiger E 1974 *Frauenrolle und Mathematikleistung*, Pädagogischer Verlag Schwann, Düsseldorf

Schmidt W, L Burstein 1985 Determinants and effects of classroom environments. Report to National Center for Education Statistics, United States, University of Illinois at Urbana-Champaign, Urbana, Il. (mimeo)

Schuard H B 1982 Differences in mathematics performance between girls and boys. In: *Mathematics Counts*. Report of the Committee of Inquiry into the Teaching of Mathematics in Schools. Her Majesty's Stationery Office, London, pp. 373–87

Servais W 1975 Continental traditions and reforms. *Int. J. Math. Educ. Sci. Technol.* 6(1) 37–58

Statistical History of the United States from Colonial Times to the Present 1976 Basic Books, New York

Steiner H G 1980 *Comparative Studies of Mathematics Curricula: Change and Stability 1960–1980*, Materialen und Studien Band 19, Institut für Didaktik der Mathematik der Universität Bielefeld, Bielefeld, FRG

Steinkamp M W, D L Harnisch, H J Walberg, S–L Tsai 1985 Cross-national gender differences in mathematics attitude and achievement among 13-year-olds. *J. Mathematical Behavior* 4: 259–77

Thurow L C 1985 *The Zero-Sum Society: Building a World-Class American Economy*. Simon and Schuster, New York

UNESCO 1981 *Statistical Yearbook 1981*. UNESCO, Paris

Usiskin Z 1986/87 The UCSMP: Translating grades 7–12 mathematics recommendations into reality. *Educational Leadership* 44(4): 30–35

Van der Blij F, S Hilding, A I Weinzweig 1980 A synthesis of national reports on changes in curricula. In: Steiner H G (ed) *Comparative Studies of Mathematics Curricula: Change and Stability 1960–1980*. Materialen und Studien Band 19, Institut für Didaktik der Mathematik der Universität Bielefeld, Bielefeld, FRG, pp. 37–54

Van der Flier H 1977 Environmental factors and deviant response patterns. In: Poortinga Y H (ed) 1977 *Basic Problems in Cross-Cultural Psychology*. Swets and Seitlinger B V, Amsterdam

Van der Flier H 1982 Deviant response patterns and comparability of test scores. *Journal of Cross-cultural Psychology*. 13: 267–298

Wattanawaha N 1986 *A Study of Equity in Mathematics Teaching and Learning in Lower Secondary School in Thailand*. Doctoral dissertation, University of Illinois at Urbana-Champaign

Westbury I, N Wattanawaha 1987 Mathematics Achievement: A comparison among middle- and high-income nations. *Pacific Education*. 1:23–39

Whitehead A N 1929 *The Aims of Education*. MacMillan, New York

Wilson J W 1971 Evaluation of learning in secondary school mathematics. In: Bloom B S, J T Hastings, and G F Madaus (eds) 1971 *Handbook on Formative and Summative Evaluation of Student Learning* McGraw-Hill, New York, pp. 643–696

Wilson James W 1976 A Second International Study of Achievement in Mathematics. University of Georgia, Athens, United States (mimeo)

Wojciechowska A 1986 *Change and failure in mathematics curricula*. Mathematical Institute, University of Wroclaw

National Reports from the Second International Mathematics Study

Australia

Rosier M J 1980 *Changes in Secondary School Mathematics in Australia 1964–1978* ACER Research Monograph No. 8. Australian Council for Educational Research, Hawthorn, Victoria, Australia

Rosier M J 1980 *Sampling Administration and Data Collection for the Second IEA Mathematics Study in Australia*. Australian Council for Educational Research, Hawthorn, Australia

Rosier M J 1980 *Test and Item Statistics for the First and Second IEA Mathematics Studies in Australia*. Australian Council for Educational Research, Hawthorn, Australia

Canada (British Columbia)

Robitaille David F, J Thomas O'Shea, Michael Dirks 1982 *The Teaching and Learning of Mathematics in British Columbia*. British Columbia Ministry of Education, Learning Assessment Branch, Victoria BC

Robitaille D F 1985 *An Analysis of Selected Achievement Data from the Second International Mathematics Study*. Victoria, British Columbia: Ministry of Education, Research and Assessment Branch, Victoria BC

Canada (Ontario)

McLean L, D Raphael, M Wahlstrom 1986 *Intentions and Attainments in the Teaching and Learning of Mathematics: Report on the Second International Mathematics Study in Ontario, Canada*. Ontario Ministry of Education, Toronto

McLean L, R Wolfe, M Wahlstrom 1987 *Learning About Teaching from Comparative Studies: Ontario Mathematics in International Perspective*. Ontario Ministry of Education, Toronto

Raphael D, M Wahlstrom, L McLean 1983 *The Second International Study of Mathematics: An Overview of the Ontario Grade 8 Study*. Ontario Institute for Studies in Education, Toronto

Raphael D, M Wahlstrom, L McLean 1983 *The Second International Study of Mathematics: An Overview of the Ontario Grade 12/13 Study*. Ontario Institute for Studies in Education, Toronto

Finland

Kangasniemi E 1988 *Opetussuunnitelma ja matematiikan koulusaavutukset*. (Curriculum and student achievement in mathematics). Institute for Educational Research. Publication series A. Research reports, University of Jyväskylä, Finland

France

Robin D, E Barrier 1985 *Enquête internationale sur l'enseignement des mathématiques: Le cas français* (International Mathematics Study: The French Case) Tome 1. INPR, Collection. Institut National de Récherche Pédagogique. "Rapports de recherches" 8. Paris

Hong Kong

Brimer A, P Griffin 1985 *Mathematics Achievement in Hong Kong Secondary Schools*. Center for Asian Studies, University of Hong Kong, Hong Kong

Japan

Sawada T, Ed 1981 *Mathematics Achievement of Secondary School Students*. National Institute for Educational Research, Tokyo. [In Japanese]

Sawada T, Ed 1982 *Mathematics Achievement and Associated Factors of Secondary School Students*. National Institute for Educational Research, Tokyo. [In Japanese]

228   The IEA Study of Mathematics

Sawada T, Ed 1983 *Mathematics Achievement and Teaching Practice in Lower Secondary Schools* (Grade 7). National Institute for Educational Research, Tokyo. [In Japanese]

Netherlands

Kuper J, W J Pelgrum 1983 *Het IEA Tweede Wiskunde Project: Aspecten van Meetkunde* [The IEA Second Mathematics Study: Aspects of Geometry]. Universiteit Twente, Toegepaste Onderwijskunde, Enschede

Pelgrum W J, Th J H M Eggen, Tj Plomp 1983 *Het IEA Tweede Wiskunde Project: Beschrijving van uitkomsten* [The IEA Second Mathematics Study: Description of Results]. Universiteit Twente. Toegepaste Onderwijskunde, Enschede

Pelgrum W J, Th J H M Eggen 1983 *Het IEA Tweede Wiskunde Project: Opzet en uitvoering* [The IEA Second Mathematics Study: Design and Execution]. Universiteit Twente, Toegepaste Onderwijskunde, Enschede

Pelgrum W J, Th J H M Eggen, Tj Plomp 1983 *Het IEA Tweede Wiskunde Project: Analyse van uitkomsten* [The IEA Second Mathematics Study: Analysis of Results]. Universiteit Twente, Toegepaste Onderwijskunde, Enschede

New Zealand

Mathematics Achievement in New Zealand Secondary Schools 1987. Department of Education, Wellington, New Zealand

Sweden

Ministry of Education 1986 *Matematik i Skolan: Översyn av undervisningen i matematik inom skolväsendet* [Mathematics in the School: Examining the Teaching of Mathematics in the School System]. Ministry of Education and Liber Allmänna Förlaget, Stockholm, Sweden

Ministry of Education 1986 *Matematik för Alla: ett diskussionsmaterial* [Mathematics for All: Material for Discussion]. Ministry of Education and Liber Utbilningsförlget, Stockholm, Sweden

Ministry of Education 1987 *Namnaren*. 86/87 (2–3): 1–12B

Murray A, R Liljefors 1983 *Matematik i Svensk Skola*. Skoloverstyrelsen

Thailand

Thai National Committee for the Second International Mathematics Study 1984 *The Analysis of the Mathematics Curriculum*. The Institute for the Promotion of Teaching Science and Technology, Bangkok. [In Thai]

Thai National Committee for the Second International Mathematics Study 1985 *The Evaluation of Mathematics Teaching and Learning in Mattayom Suksa 5 Classes*. The Institute for the Promotion of Teaching Science and Technology, Bangkok. [In Thai]

Thai National Committee for the Second International Mathematics Study. 1986 *The Evaluation of Mathematics Teaching and Learning in Mattayom Suksa 2 Classes*. The Institute for the Promotion of Teaching Science and Technology, Bangkok. [In Thai]

United States

Chang A et al. 1985 *Technical Report I, Item Level Achievement and OTL Data*. Stipes Publishing Company, Champaign, IL

Crosswhite J et al. 1985 *United States Summary Report: Second International Mathematics Study*. Stipes Publishing Company, Champaign, IL

Crosswhite J et al. 1986 *Detailed National Report*. Stipes Publishing Company, Champaign, IL
McKnight C et al. 1987 *The Underachieving Curriculum: Assessing U.S. School Mathematics from an International Perspective* Stipes Publishing Company, Champaign, IL
United States Summary Report: Second International Mathematics Study. Stipes Publishing Company, Champaign, IL
Technical Report I, Item Level Achievement and OTL Data Stipes Publishing Company, Champaign, IL

International Reports from the Second International Mathematics Study

Dossey J A et al. 1987 Mathematics teachers in the United States and other countries participating in the Second International Mathematics Study. University of Illinois at Urbana-Champaign, Urbana, IL
Garden R A 1987 *Second IEA Mathematics Study Sampling Report*. Center for Education Statistics, U.S. Department of Education, Washington, D.C., USA
Jaji G 1986 The uses of calculators and computers in mathematics classes in twenty countries: Summary Report. University of Illinois at Urbana-Champaign, Urbana, IL
Kifer E, R G Wolfe, W H Schmidt 1985 Cognitive Growth in Eight Countries: Lower Secondary School. University of Illinois at Urbana-Champaign, Urbana, IL
Livingstone I D 1986 *Perceptions of the Intended and Implemented Mathematics Curriculum* Center for Education Statistics. U.S. Department of Education, Washington, D.C., USA
Miller D M, R L Linn 1986 Cross-National achievement with differential retention rates. University of Illinois at Urbana-Champaign, Urbana, IL
Oldham E E 1986 Qualification of mathematics teachers. Trinity College, Dublin, Ireland
Pelgrum W J, Th J H M Eggen, T J Plomp 1985 The implemented and attained mathematics curriculum: A comparison of twenty countries. Twente University of Technology, Enschede, The Netherlands
Robitaille D F, A R Taylor 1985 A comparative review of students' achievement in the First and Second IEA Mathematics Studies. University of British Columbia, Vancouver, Canada
Wolfe R G 1987 A comparative analysis across eight countries of growth in mathematics achievement: The effects of one year of instruction. Ontario Institute for Studies in Education, Toronto, Canada
Wolfe R G 1986 *Training Manual for Use of the Data Bank of the Longitudinal, Classroom Process Surveys for Population A in the IEA Second International Mathematics Study*. Ontario Institute for Studies in Education, Toronto, Canada
Wolfe R G, D Raphael 1985 Reports of Homework Assignment Around the World: Results from the Second International Mathematics Study. Ontario Institute for Studies in Education, Toronto, Canada

APPENDIX I

Participating Systems

System	National Research Coordinator	Council Member	National Center
Australia*†	Malcolm Rosier	John Keeves	Australian Council for Educational Research, Hawthorn, Victoria
Belgium (Flemish)	Christiana Brussel-mans-Dehairs	A. De Block (1976– 81) J. Heene	Seminaire en Laboratorium voor Didactiek, Gent
Belgium* (French)	Georges Henry	Gilbert De Landsheere	Université de Liege au Sart Tilman
Canada (B.C.)	David Robitaille	Joyce Matheson	Ministry of Education, Victoria
England/Wales*	Michael Cresswell (1978–83) Derek Foxman	Clare Burstall	National Foundation for Educational Research, Slough
Finland*	Erkki Kangasniemi	Kimmo Leimu	Institute for Educational Research, University of Jyväskylä, Jyväskylä
France*	Daniel Robin	Daniel Robin	Institut National de Recherche Pédagogique, Paris
Hong Kong	Patrick Griffin (1976–1983)	M. A. Brimer	University of Hong Kong, Hong Kong
Hungary	Julia Szendrei	Zoltán Báthory	Országos Pedagógiai Intézet, Budapest
Ireland†	Elizabeth Oldham	Elizabeth Oldham	Trinity College, Dublin
Israel*	Arieh Lewy (1976–83) David Nevo	Arieh Lewy (1976–83) David Nevo	University of Tel-Aviv Tel-Aviv
Ivory Coast†	Sango Djibril	Ignace Koffi	Service D'Evaluation, Abidjan
Japan*	Toshio Sawada	Hiroshi Kida	National Institute for Educational Research, Tokyo
Luxembourg	Robert Dieschbourg	Robert Dieschbourg	Institut Pédagogique, Walferdange
Netherlands*	Tjeerd Plomp (1976–1979) Hans Pelgrum	Egbert Warries	Twente University of Technology, Enschede
New Zealand	Robert Garden Athol Binns	Roy W. Phillipps	Department of Education, Wellington
Nigeria	Wole Falayajo	E. A. Yoloye	University of Ibadan, Ibadan

Scotland*	Gerard Pollock	W. Bryan Dockrell	Scottish Council for Research in Education, Edinburgh
Swaziland	Mats Eklund (1976–1981) P. Simelane	Mats Eklund (1976–1981) P. Simelane	William Pitcher College, Manzini
Sweden*	Robert Liljefors	Torsten Husén (1976–83) Inger Marklund	University of Stockholm, Stockholm
Thailand	Samrerng Boonruangrutana	Pote Sapianchai	Institute for the Promotion of Teaching Science and Technology, Bangkok
United States*	Curtis McKnight	Richard Wolf	Teachers College, Columbia University New York

*These systems, together with the Federal Republic of Germany, participated in the First IEA Study of Mathematics.

†Did not participate in testing phases of the Study. Australia and Ireland collected data independently. See Rosier (1980).

APPENDIX 2

Timeline for the Second International Mathematics Study

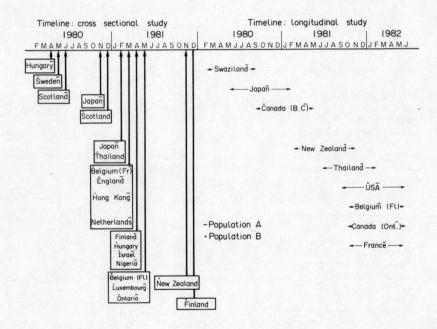

Index